Introducing Design Automation for Quantum Computing

Alwin Zulehner • Robert Wille

Introducing Design Automation for Quantum Computing

 Springer

Alwin Zulehner
Johannes Kepler University of Linz
Linz, Austria

Robert Wille
Johannes Kepler University of Linz
Linz, Austria

ISBN 978-3-030-41755-0   ISBN 978-3-030-41753-6   (eBook)
https://doi.org/10.1007/978-3-030-41753-6

This Springer imprint is published by the registered company Springer Nature Switzerland AG.
The registered company address is: Gewerbestrasse 11, 6330 Cham, Switzerland

Preface

In the 1970s, researchers started to utilize quantum mechanics to address questions in computer science and information theory—establishing new research directions (such as quantum computing) where quantum-mechanical effects allow for a substantial improvement of information density and computation power. Now, more than four decades later, we are at the dawn of a new "computing age" in which quantum computers indeed will find their way into practical application. Besides specialized startups like *Rigetti* and *IonQ*, also big players like *IBM*, *Google*, *Microsoft*, and *Intel* have entered the field and, e.g., provide physical implementations of quantum computers that are made publicly available through cloud services.

While impressive accomplishments are observed in the physical realization of quantum computers, the development of automated tools and methods that provide assistance in the design and realization of applications for those devices is at risk of not being able to keep up with this development anymore—leaving a situation where we might have powerful quantum computers but hardly any proper means to actually use them. This is different from the conventional realm (i.e., for electronic circuits and system), where decades of research led to sophisticated design tools for software and hardware that can handle complexity of impressive scales—a main reason for the penetration of electronic devices into almost all parts of our daily life.

This book provides the foundation for similar accomplishments in the quantum realm. It introduces expertise on efficient data structures and algorithms gained in the design automation community over the last decades and shows how this can be utilized for several design tasks relevant for quantum computing. This results in a variety of methods and tools tackling the challenges in quantum computing from a design automation perspective. More precisely, this book presents efficient solutions for

- quantum-circuit simulation,
- the design of Boolean components needed in quantum circuits, and
- the mapping of quantum circuits to real hardware (also known as compilation).

Evaluations show that, using the proposed solutions, several orders of magnitude with respect to run-time or other design objectives are frequently gained—showcasing the tremendous further potential of design automation for quantum computing.

By this, this book significantly contributes to the recently started (and highly demanded) trend of bringing knowledge from the design automation community to the quantum realm. The resulting methods and algorithms led to several efficient tools that are publicly available as open-source implementations. Some of them are even officially integrated (or planned to) into IBM's SDK Qiskit or Atos' Quantum Learning Machine—making them directly available for anyone using IBM's quantum computers or Atos' quantum programming language.

Overall, this book is the result of several years of research conducted at the Johannes Kepler University Linz, Austria, as well as numerous discussions with colleagues and stakeholders from around the world. For this, we would like to particularly thank Lukas Burgholzer, Stefan Hillmich, and Armando Rastelli from our home university, Yehuda Naveh and the team of IBM Research in Haifa, Rudy Raymond and the team of IBM Research in Tokyo, Ali Javadi-Abhari and Andrew Cross from the IBM Thomas J. Watson Research Center, Austin Fowler and Sergio Boixo from Google Inc., Philipp Niemann and Rolf Drechsler from the University of Bremen, Igor L. Markov from the University of Michigan, Carmina G. Almudever and the team of TU Delft, Cyril Allouche and his team from Atos, Martin Roetteller and his team from Microsoft Research, as well as many more. Besides that, we sincerely thank the coauthors of all the papers that formed the basis of this book. Furthermore, we would like to thank the Institute for Integrated Circuits as well as the LIT Secure and Correct Systems Lab at the Johannes Kepler University Linz for always providing a stimulating and enjoyable environment in which scientific ideas indeed can flourish. For funding, we thank the State of Upper Austria (within the context of the LIT Secure and Correct Systems Lab) as well as the European Union (within the context of the COST Action IC1405). Finally, we would like to thank Springer Nature and especially Charles "Chuck" Glaser for publishing this work.

Linz, Austria   Alwin Zulehner
December 2019   Robert Wille

Contents

Part I
Introduction and Background

Chapter 1
Introduction

Abstract This chapter sets the context of this book by providing an overview on the considered design tasks in quantum computing, their importance, as well as their complexity. Based on that, the importance of bringing knowledge (on clever data structures and efficient algorithms) from the design automation community into the quantum realm is discussed—to avoid a situation, where we might end up with powerful quantum hardware but without efficient tools to make full use of that power.

Keywords Quantum computer · Quantum circuits · Qubits · Design automation · Complexity

In the 1970s, researchers started to utilize quantum mechanics to address questions in computer science and information theory—establishing new research directions such as quantum computing, quantum information, or quantum security [114]. In all these fields, the so-called quantum bits (i.e., qubits) serve as elementary information unit, which—in contrast to conventional bits—can not only be in one of its two orthogonal basis states (denoted $|0\rangle$ and $|1\rangle$ using Dirac notation), but also in an almost arbitrary superposition of both (i.e., $\alpha_0 \cdot |0\rangle + \alpha_1 \cdot |1\rangle$, where the complex factors α_0 and α_1 satisfy $\alpha_0\alpha_0^* + \alpha_1\alpha_1^* = 1$). This allows an n-qubit quantum system to represent 2^n different complex values at once—exponentially more than conventional n-bit systems (which only can represent n different Boolean values at a time). Together with further quantum-physical phenomena such as entanglement [114] (the state of a qubit might be influenced by the state of other qubits), this allows for substantial improvements in information density as well as computation power, and motivated the establishment of dedicated research areas investigating and exploiting this potential.

One of these research areas, namely quantum computing, covers the development of applications that exploit these quantum-physical effects in order to solve certain problems significantly faster (in the best case, exponentially faster) than conventional computers. First corresponding quantum algorithms such as Grover Search or Shor's Algorithm, which improve database search [59] and

© Springer Nature Switzerland AG 2020
A. Zulehner, R. Wille, *Introducing Design Automation for Quantum Computing*,
https://doi.org/10.1007/978-3-030-41753-6_1

enable integer factorization in polynomial time [143], respectively, have already been proposed in the last century. Since then, corresponding developments have significantly broadened and, in the meantime, cover a huge variety of quantum applications including quantum chemistry, solving systems of linear equations, physics simulations, machine learning, and many more [30, 111, 126]. In this regard, quantum chemistry, which allows simulating molecules, e.g., to find better catalysts for the Haber–Bosch process (which is consuming 1–2 % of the world's energy supply and savings would have a tremendous impact), are considered very promising to be useful in the near future.

These developments are also triggered by the fact that quantum computers are reaching feasibility.[1] In fact, the importance and potential impact of this disruptively new computation paradigm did not only led academia to work on corresponding first physical realizations (presented, e.g., in [40, 92, 112]). Nowadays, also "big players" such as *IBM, Google, Microsoft*, and *Intel* as well as specialized startups such as *Rigetti* and *IonQ* have entered this research field and are heavily investing in it [32, 52, 55, 70, 81, 85, 135]. In 2017, this led to the first quantum computers that are publicly available through cloud access by IBM [73]. Since then, their machines have been used by more than 100,000 users, who have run more than 6.5 million experiments thus far. Recently, IBM followed with the presentation of their prototype towards a quantum computer for commercial use (a stand-alone quantum computer to be operated outside of their labs)—the *IBM Q System One* presented in January 2019 at CES [72].

Assuming that the progress in quantum-computer development (i.e., increasing the number of qubits, gate fidelity, as well as coherence time of qubits) continues as currently observed, practically relevant physical realizations that outperform conventional ones (showing *quantum supremacy* [16]) are within reach—at least for some of the applications mentioned above [30, 111, 126].[2] However, there is even more potential for quantum computing that can be unveiled by turning the currently developed *Noisy Intermediate-Scale Quantum* (NISQ [126]) devices into *fault-tolerant* ones that are capable of conducting very deep computations on a large number of qubits and with perfect accuracy—allowing to solve even more problems in an efficient fashion. The key elements for such fault-tolerant devices are—besides further reduction of error rates and improvement of coherence time— error-correcting codes where each logical qubit in a computation is realized by several (up to several hundreds) of physical qubits [12, 50, 56, 69, 125].

However, despite these accomplishments and prospects, also the development of automated tools and methods that provide assistance in the design and realization of corresponding applications is required. Otherwise, we are approaching a situation

---

[1]Note that quantum computers require conventional computers to run/control those [51]. Hence, they will, presumably, always serve as an accelerator for conventional machines (for certain problems) rather than replacing them completely.

[2]Researches from Google have recently claimed that they achieved quantum supremacy on their Sycamore quantum computer [8]—a claim that might be over-ambitious as discussed in a recent work by IBM [123] but certainly shows the progress achieved thus far.

where we might have powerful quantum computers, but hardly any proper means to actually use them. Among others, the following design tasks—described using the model of *quantum circuits*[3]—are of importance in this regard:

- **Simulation** of quantum circuits on conventional computers plays an important role. Since real quantum computers are expensive, hardly available, and yet limited in the number of qubits as well as their fidelity, simulators serve as a temporary substitute that allows designing quantum applications, as well as to evaluate error-correcting codes already today. Additionally, given the extreme difficulty of formal verification for quantum computation [1, 16, 34], simulation can be adapted to circuit equivalence-checking and other functional verification tasks [118, 162, 167, 181]. In both scenarios, simulators may give additional insights since, e.g., the precise amplitudes of a quantum state are explicitly determined (while they are not observable in a real quantum computer). Finally, simulators serve as a performance baseline for quantifying advantages of quantum computers [8, 123], e.g., efficient simulation of certain circuits inherently indicates that no significant quantum speedup is reachable in these cases.

  However, quantum-circuit simulation in general constitutes a computationally very complex task since each quantum gate and each quantum state are eventually represented by a unitary matrix or state vector that grows exponentially with the number of qubits. In fact, each quantum operation applied to a quantum state composed of n qubits requires multiplying a $2^n \times 2^n$-dimensional matrix with a 2^n-dimensional vector.[4] This constitutes a serious bottleneck, which prevents the simulation of many quantum applications and, by this, the evaluation of their potential.

- **Compilation** of quantum algorithms containing high-level operations (e.g., modular exponentiation in Shor's algorithm) into elementary gates available on the target architecture (which, e.g., directly correspond to microwave impulses that are applied to the superconducting transmon qubits in IBM's or Google's devices) is required to run them on real quantum hardware. Thereby, circuits composed of gates that operate on a single qubit and may be controlled by further qubits usually serve as intermediate description means in the compilation process. Despite the terminology correlation to physical (i.e., hardware) gates known from the conventional realm, a quantum gate describes an operation that is applied to a certain set of qubits rather than a physical entity. Hence, a quantum circuit actually represents a program to be executed on quantum hardware—

---

[3]Since quantum circuits map well to leading quantum-computing platforms, serve as description means for quantum algorithms, and also can simulate other (digital or analog) models of quantum computation with modest overhead [5, 27, 110, 127], methods presented in this book focus on quantum circuits without the loss of the generality.

[4]Note that different simulation approaches exist that do not compute the complete final state vector, and that it is usually not necessary to represent the exponentially large matrix explicitly. However, this does not decrease the exponential complexity.

making the terminology of compilation appropriate. In order to compile quantum algorithms, automated methods for the following tasks are required [150, 151]:

- *Design of Boolean Components*: For the "quantum part" of an algorithm, the decomposition into multiple-controlled one-qubit gates is usually inherently given by the algorithm, by using common building blocks (e.g., QFT), or determined by hand.[5] This is different for the needed Boolean components (e.g., arithmetic components), which often employ a rather complex functionality. Since quantum circuits are inherently reversible [114] and only Boolean functionality is realized, automated approaches for *reversible-circuit synthesis* are required[6]—a very challenging task due to its exponential complexity [155].

- *Mapping to NISQ Devices*: To run quantum circuits on real hardware, they have to be mapped to the target architecture. In a first step, the gates of the circuit to be mapped are decomposed into elementary operations that are available on the target hardware or in a certain gate library (e.g., *Clifford+T* [19]). Usually, this gate library contains a single two-qubit gate in conjunction with a variety of one-qubit gates—allowing for universal quantum computing. Then, the logical qubits of a quantum circuit are mapped to the physical ones of a quantum computer—a non-trivial task since connectivity limitations given by the target architecture prohibit interactions between certain pairs of qubits. This is usually resolved by inserting further operations that change the mapping of the qubits dynamically—yielding an $\mathcal{NP}$-complete task.

In the conventional realm (i.e., for electronic circuits and systems), similar design tasks are well-defined and have been studied by researchers and engineers for decades—resulting in sophisticated design tools for software as well as for hardware that are taken for granted today and that can handle complexity of impressive scales. The availability of these methods and tools is a main reason for the utilization and penetration of (conventional) electronic devices into almost all parts of our daily life.

However, for the quantum realm, the corresponding state of the art is (yet) completely different. Indeed, active development is present (as, e.g., witnessed by recent launches of development environments by "big players" such as IBM's *Qiskit* [33], Google's *Cirq* [28], Microsoft's *Quantum Development Kit* [101], or Rigetti's *Forest* [130]), but none of these toolkits provides a collection of design methods that is as powerful as approaches available for the conventional realm, nor do they cover a similarly broad variety of design tasks. Moreover, they often do not (yet) utilize experience and expertise gained in the design automation community

---

[5]Note that also few automated approaches exist [116, 117, 119].

[6]Note that the term synthesis is appropriate in the context of reversible circuits (which are utilized to realize Boolean components) since they also have applications in different areas (besides quantum computation) where they might eventually be fabricated as physical circuits in conventional hardware [9, 128, 194].

over the last decades—even though, as discussed above, many of the considered design problems are similar since they are of combinatorial and exponential nature (some have even been proven to be $\mathcal{NP}$-complete [18, 144] or co$\mathcal{NP}$-hard [155]).

This book provides the foundation to change that by introducing additional expertise on clever as well as efficient data structures and algorithms (gained over the last decades in the design automation community for the conventional realm) to the area of quantum computing. To this end, the relevant design tasks outlined above—*quantum-circuit simulation*, the *design of Boolean components*, as well as *quantum-circuit mapping*—are carefully investigated from a design automation perspective in self-contained parts of this book (i.e., Part II, Part III, and Part IV, respectively)—yielding complementary and efficient approaches that are inspired by core techniques known from design automation for conventional circuits and systems. More precisely, relying on, e.g., accordingly adapted versions of decision diagrams, informed search, or powerful reasoning engines allows providing highly efficient complementary approaches that significantly outperform the current state of the art by orders of magnitude (with respect to run-time or corresponding design objectives). All these developments further contribute to the envisioned goal and showcase the tremendous available potential—eventually leading to a set of sophisticated tools that enhance the development of quantum computing itself, and leverage its way from an initially academically driven dream to fully fletched system following this powerful computation paradigm.

Chapter 2
Quantum Computing

Abstract This chapter provides a background on quantum computing, quantum and reversible circuits, as well as on currently developed quantum hardware (i.e., NISQ devices). In this context, also the notation and naming conventions used throughout this book are introduced.

Keywords Quantum computer · Quantum gates · Qubits · Quantum state · Quantum circuit · IBM Q

To keep this book self-contained, this chapter provides the basics of quantum computing, i.e., it reviews quantum states and operations, currently available quantum technology, as well as reversible and quantum circuits.

2.1 Quantum States and Operations

This book focuses on models following digital quantum computation. They are characterized by discrete-time steps (e.g., *quantum Turing machines* [41], *quantum circuits* [110, 182], or *one-way quantum computation* [129]), whereas models following analog quantum computation (mainly the *adiabatic model of computation* [5, 47], which, e.g., is utilized in devices developed by *D-Wave* [83]) perform *continuous-time information processing*. Conventional digital logic operates on bits, i.e., binary digits that can be concatenated to express larger values. In contrast, quantum computations operate on *qubits*—two-level quantum systems that can be combined into n-qubit systems. Each qubit is typically initialized in one particular state labeled $|0\rangle$, and different states such as $|1\rangle$ are obtained by modifying the qubit. As long as $|0\rangle$ and $|1\rangle$ are orthogonal, one can use them as the computational basis. Then, the state of a qubit is given by a linear combination (i.e., a *superposition*) of these basis states $|\varphi\rangle = \alpha_0 \cdot |0\rangle + \alpha_1 \cdot |1\rangle$, where the complex *amplitudes* α_0 and α_1 satisfy $\alpha_0\alpha_0^* + \alpha_1\alpha_1^* = 1$. The *Bloch Sphere* (depicted in Fig. 2.1) gives a geometric interpretation of the state of a qubit. Here, the basis states $|0\rangle$ and $|1\rangle$ are located at

© Springer Nature Switzerland AG 2020
A. Zulehner, R. Wille, *Introducing Design Automation for Quantum Computing*,
https://doi.org/10.1007/978-3-030-41753-6_2

Fig. 2.1 Bloch sphere

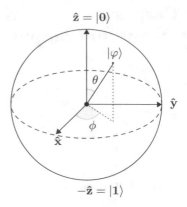

the north pole and south pole of the sphere, respectively, whereas possible quantum states are represented by points on the surface of the sphere, by specifying angles θ and ϕ such that $|\varphi\rangle = \cos\left(\frac{\theta}{2}\right) \cdot |0\rangle + e^{i\phi} \sin\left(\frac{\theta}{2}\right) \cdot |1\rangle$.

Physical realizations of qubits vary, but when working with spin-½ particles, $|0\rangle$ is typically assigned to the ground (least-energy) state, and some form of cooling performs qubit initialization. Qubit states different from $|0\rangle$ gradually decay (*decohere*) to the ground state, which limits the amount of computation that can be performed.

The composition of qubits into a larger system is modeled by the *tensor product* operation that generalizes the concatenation of conventional bits. In particular, the joint state of n qubits (also denoted as the system's *wave function*) is contained in the *tensor product* of n two-dimensional Hilbert spaces—the 2^n-dimensional Hilbert space spanned by the basis $|0\rangle, \ldots, |2^n - 1\rangle$. Writing out these integer labels in binary makes the correspondence between the two interpretations explicit, e.g., for three qubits, the basis vector $|6\rangle = |110_2\rangle = |1\rangle\otimes|1\rangle\otimes|0\rangle$ represents a configuration with the least significant qubit assuming $|0\rangle$ and the other two qubits assuming $|1\rangle$. While the 2^n computational-basis states are specified by bitstrings of length n, their *superposition* (linear combination) may need up to 2^n complex-valued parameters— appearing as the amplitudes of the unit-norm state vector.

Definition 2.1 *Consider a quantum system composed of n qubits. Then, all possible states of the system are of the form*

$$|\varphi\rangle = \sum_{x\in\{0,1\}^n} \alpha_x \cdot |x\rangle, \; \text{where} \; \sum_{x\in\{0,1\}^n} \alpha_x\alpha_x^* = 1 \; \text{and} \; \alpha_x \in \mathbb{C}.$$

The state $|\varphi\rangle$ *can be also represented by a column vector* $\varphi = [\varphi_i]$ *with* $0 \le i < 2^n$ *and* $\varphi_i = \alpha_x$*, where* $nat(x) = i$.[1]

---

[1]Note that, to save space, vectors may be provided in their transposed form in the following (indicated by $[\cdot]^T$). That is, the single elements are listed horizontally rather than vertically.

Some superpositions are *separable* and can be compactly represented as tensor products of smaller states, e.g., $|\varphi\rangle = (|01\rangle + |11\rangle)/\sqrt{2} = (|0\rangle + |1\rangle)/\sqrt{2} \otimes |1\rangle$. Yet, most states cannot be represented this way and are termed *entangled*. Then, the state of a qubit depends on the state of other qubits (i.e., they influence each other), e.g., $|\varphi\rangle = (|00\rangle + |11\rangle)/\sqrt{2}$. Together with superposition, entanglement provides the basis for substantial improvements in information density as well as computation power compared to the conventional realm.

Quantum states cannot be directly observed. To extract (partial) information from quantum states in the form of conventional bits, one performs a *measurement opera-tion*. In contrast to conventional computers, this measurement modifies the quantum state.[2] The possible outcomes of a *measurement operation* form an (orthogonal) basis. In the process of measurement, the quantum state non-deterministically collapses to one of these basis states where the probability of each outcome reflects the proximity to the respective basis state. More precisely, measuring a one-qubit state $\alpha_0 \cdot |0\rangle + \alpha_1 \cdot |1\rangle$ (with $\alpha_0\alpha_0^* + \alpha_1\alpha_1^* = 1$) in the computational basis ($|0\rangle$ and $|1\rangle$) changes the state to $|0\rangle$ or $|1\rangle$ with probabilities $\alpha_0\alpha_0^*$ and $\alpha_1\alpha_1^*$, respectively.[3] A measurement of an *n*-qubit state in the computational basis can be physically implemented by measuring every qubit in its computational basis. Here, the amplitude α_k of the state determines the probability $\alpha_k\alpha_k^*$ of outcome $|k\rangle$, where the binary form of k specifies the outcomes of individual qubits. When some qubits are left unmeasured, the probability of a particular measurement outcome is a sum over the probabilities of all (would-be) compatible full-measurement outcomes. Note that computing the probability from an amplitude neglects the *quantum phase* (the angular component of a complex number), which is why the *global phase* of a quantum state is unobservable.

Example 2.1 *Consider a quantum system composed of n = 3 qubits q_0, q_1, and q_2 that assumes the state $|\varphi\rangle = |q_0q_1q_2\rangle = \frac{1}{2} \cdot |010\rangle + \frac{1}{2} \cdot |100\rangle - \frac{1}{\sqrt{2}} \cdot |110\rangle$. Then, the state vector of the system is given by*

$$\varphi = \left[0, 0, \frac{1}{2}, 0, \frac{1}{2}, 0, -\frac{1}{\sqrt{2}}, 0\right]^T.$$

Measuring the system yields basis states $|010\rangle$, $|100\rangle$, and $|110\rangle$ with probabilities $\frac{1}{4}$, $\frac{1}{4}$, and $\frac{1}{2}$, respectively. Measuring only qubit q_0 collapses q_0 into basis state $|0\rangle$ and $|1\rangle$ with probabilities $\frac{1}{4}$ and $\frac{1}{4} + \frac{1}{2} = \frac{3}{4}$, respectively, changing the state of the system either to $|\varphi'\rangle = |010\rangle$ or to $|\varphi''\rangle = \frac{1}{\sqrt{3}} \cdot |100\rangle - \sqrt{\frac{2}{3}} \cdot |110\rangle$.

Aside from measurements, quantum computers apply quantum operations to a fixed set of qubits, altering the joint state of the qubits.

---

[2] *Weak measurements* [87] are not considered in this book.

[3] A different basis can be specified by the eigenvectors of a Hermitian ($H = H^*$) matrix, such as the Pauli X, Y, Z matrices defined below.

Definition 2.2 *Consider a quantum system composed of n qubits. Each quantum operation acting on this state is expressed as $2^n \times 2^n$ unitary matrix U (i.e., a matrix satisfying $UU^* = I$) and, hence, is inherently reversible.*

However, simple quantum operations (also denoted *gates*) are defined over one or two qubits only and specified by (1) a small matrix as well as (2) the qubits on which the gate is applied. Mathematically speaking, the resulting $2^n \times 2^n$ matrix can then be computed as the Kronecker product of the gate matrix and a large identity matrix.

The state of an individual qubit can be modified by a one-qubit gate that is represented by a 2×2 unitary matrix in the computational basis. Commonly used quantum gates for generating a superposition (the Hadamard operation H), inverting a quantum state (X), and applying phase shifts by -1 (Z), i (P), and $\frac{1+i}{\sqrt{2}}$ (T) are, respectively, defined as

$$H = \frac{1}{\sqrt{2}}\begin{bmatrix} 1 & 1 \\ 1 & -1 \end{bmatrix}, \text{NOT} = X = \begin{bmatrix} 0 & 1 \\ 1 & 0 \end{bmatrix}, \quad Y = \begin{bmatrix} 0 & -i \\ i & 0 \end{bmatrix},$$

$$Z = \begin{bmatrix} 1 & 0 \\ 0 & -1 \end{bmatrix}, \quad P = S = \begin{bmatrix} 1 & 0 \\ 0 & i \end{bmatrix}, \text{T} = \begin{bmatrix} 1 & 0 \\ 0 & \frac{1+i}{\sqrt{2}} \end{bmatrix}.$$

One-qubit gates are often defined as rotations on the *Bloch sphere* (cf. Fig. 2.1). Rotations around the axes of the sphere are written as

$$R_x(\theta) = \begin{bmatrix} \cos\frac{\theta}{2} & -i\sin\frac{\theta}{2} \\ -i\sin\frac{\theta}{2} & \cos\frac{\theta}{2} \end{bmatrix},$$

$$R_y(\theta) = \begin{bmatrix} \cos\frac{\theta}{2} & -\sin\frac{\theta}{2} \\ \sin\frac{\theta}{2} & \cos\frac{\theta}{2} \end{bmatrix}, \text{ and } R_z(\theta) = \begin{bmatrix} e^{-\theta/2} & 0 \\ 0 & e^{\theta/2} \end{bmatrix}.$$

For $\theta = \pi$, these expressions give the Pauli X, Y, and Z matrices up to global phase. Roots of X, Y, and Z can be obtained by taking fractions of π. Any one-qubit gate can be decomposed into a product $R_x(\theta_x)R_y(\theta_y)R_z(\theta_z)$ with some angular parameters θ_x, θ_y, and θ_z.[4]

Two-qubit gates can couple pairs of qubits and are represented by 4×4 unitary matrices. By applying arbitrary two-qubit gates to different pairs of qubits, it is possible to effect any 2^n-dimensional unitary, i.e., attain universal quantum computation. It is common to allow a variety of one-qubit gates but limit two-qubit gates to one of the following types:

---

[4]Other decompositions of the form $R_y(\theta_y)R_z(\theta_z)R_y(\theta'_y)$ are $R_z(\theta_z)R_y(\theta_y)R_z(\theta'_z)$ are possible— showing that only two gate types suffice to represent arbitrary one-qubit gates.

$$
\text{CNOT} = \begin{bmatrix} 1 & 0 & 0 & 0 \\ 0 & 1 & 0 & 0 \\ 0 & 0 & 0 & 1 \\ 0 & 0 & 1 & 0 \end{bmatrix}, \quad \text{CZ} = \begin{bmatrix} 1 & 0 & 0 & 0 \\ 0 & 1 & 0 & 0 \\ 0 & 0 & 1 & 0 \\ 0 & 0 & 0 & -1 \end{bmatrix}, \text{ and } \text{iSWAP} = \begin{bmatrix} 1 & 0 & 0 & 0 \\ 0 & 0 & i & 0 \\ 0 & i & 0 & 0 \\ 0 & 0 & 0 & 1 \end{bmatrix}.
$$

The two-qubit CNOT gate can be defined by its action on the four computational-basis vectors: $|x\,y\rangle \mapsto |x\,x \oplus y\rangle$, where $\oplus$ represents the *exclusive-or* (XOR) operation, the unmodified qubit x is called *control*, and the other bit is called *target*. Given that the CZ gate can be expressed using the CNOT gate and two H gates (and vice versa), the two gates share many properties, whereas iSWAP effects a stronger coupling between qubits. Using one of these gates with a sufficiently wide variety of one-qubit operations, e.g., arbitrary rotations around the y- and z-axis, still allows for universal quantum computation. Moreover, any unitary matrix can be broken down into such gates in multiple ways.

The state resulting from applying a quantum operation is determined by matrix–vector multiplication. When an n-qubit gate A is applied to an n-qubit quantum state v, the resulting state is $(A \cdot v)_i = \sum_{k=1}^{2^n} A_{i,k} v_k$. Therefore, when two gates are applied on the same qubits in sequence, the resulting operation is represented by the matrix product of gate matrices. When an m-qubit gate A and an n-qubit gate B are applied in parallel (on different qubits), the resulting operation is represented by the *Kronecker product* $A \otimes B$ of two matrices. In particular, elements of the nm-qubit gate $A \otimes B$ can be expressed as

$$
(A \otimes B)_{2^n i + j,\, 2^n k + l} = (a_{i,j} b_{k,l}) = \begin{bmatrix} a_{1,1} B & \cdots & a_{1,2^m} B \\ \vdots & \ddots & \vdots \\ a_{2^m,1} B & \cdots & a_{2^m,2^m} B \end{bmatrix}. \tag{2.1}
$$

The row (column) indices of $A \otimes B$ are formed by bitwise concatenation of row (column) indices of A and B. In a special case, gate A is applied to $k < n$ qubits without affecting other qubits, the $I_{2^{n-k}}$ identity matrix is assumed to act on $n - k$ qubits. The full $2^n \times 2^n$ matrix of this operation is $A \otimes I_{2^{n-k}}$.

Example 2.2 *Consider a two-qubit quantum system that is initially in state* $|\varphi\rangle = |q_0 q_1\rangle = |00\rangle$. *Assume that a Hadamard gate is applied to qubit q_0 and, afterwards, a CNOT gate with control qubit q_0 and target qubit q_1. Finally, the qubit q_0 is measured. The resulting state $|\varphi'\rangle$ (before measurement) is determined by multiplying the respective unitary matrices to the state vector. Since the Hadamard gate shall only affect q_0, the Kronecker product of* H *and the identity matrix I_2 is formed, i.e.,*

$$
H \otimes I_2 = \frac{1}{\sqrt{2}} \begin{bmatrix} 1 & 1 \\ 1 & -1 \end{bmatrix} \otimes \begin{bmatrix} 1 & 0 \\ 0 & 1 \end{bmatrix} = \frac{1}{\sqrt{2}} \begin{bmatrix} 1 & 0 & 1 & 0 \\ 0 & 1 & 0 & 1 \\ 1 & 0 & -1 & 0 \\ 0 & 1 & 0 & -1 \end{bmatrix}.
$$

Then, $|\varphi'\rangle$ *is determined by*

$$|\varphi'\rangle = \begin{bmatrix} 1 & 0 & 0 & 0 \\ 0 & 1 & 0 & 0 \\ 0 & 0 & 0 & 1 \\ 0 & 0 & 1 & 0 \end{bmatrix} \cdot \frac{1}{\sqrt{2}} \begin{bmatrix} 1 & 0 & 1 & 0 \\ 0 & 1 & 0 & 1 \\ 1 & 0 & -1 & 0 \\ 0 & 1 & 0 & -1 \end{bmatrix} \cdot \begin{bmatrix} 1 \\ 0 \\ 0 \\ 0 \end{bmatrix} = \frac{1}{\sqrt{2}} \begin{bmatrix} 1 \\ 0 \\ 0 \\ 1 \end{bmatrix}.$$

As can be seen, the two gates entangle the qubits q_0 *and* q_1—*generating a so-called Bell state* $|\varphi'\rangle = \frac{1}{\sqrt{2}}(|00\rangle + |11\rangle)$. *Measuring qubit* q_0 *collapses its superposition into one of the two basis states. Since* q_0 *and* q_1 *are entangled,* q_1 *collapses to the same basis state.*

Note that high-level descriptions of quantum algorithms often use larger (macro) gates that implement the *controlled-U* (C-U) pattern. Such a gate designates c *control qubits* and disjoint t *target qubits*, along with a t-qubit gate U (usually $t = 1$). Then, the gate operation is defined on computational-basis states and extended by linearity. In particular, basis states where all control qubits carry $|1\rangle$ are modified by applying U to the target qubits. All other basis states are left unchanged. Acting on $k = t + c$ qubits, such a gate is represented by $2^k \times 2^k$ unitary matrices of the form

$$\begin{bmatrix} I_{2^k - 2^t} & 0 \\ 0 & U \end{bmatrix}.$$

For example, the CNOT and CZ gates described above can be viewed as controlled operations with $t = c = 1$, where $U = NOT$ and $U = Z$, respectively. Also common are Toffoli gates (2-controlled X gates) used in arithmetic operations and controlled rotation gates used in Quantum Fourier Transforms. Applying controlled operations C-U acting on $k > 2$ qubits on a physical quantum computer is costly since this requires their decomposition into 1- and 2-qubit gates, e.g., using methods presented in [6, 11, 99, 106, 139, 180].

2.2   Current and Future Quantum Technology

Currently, there is an ongoing "race" between large companies like *IBM*, *Google*, *Microsoft*, *Intel*, *Rigetti*, and *IonQ* to build the first practically useful quantum computer [32, 52, 55, 70, 81, 85, 135]. They all develop devices classified as *Noisy Intermediate-Scale Quantum* (NISQ [126]) technology. Although still limited by their number of available qubits and low fidelity, NISQ devices (of sufficient size and fidelity) provide the capability of running quantum algorithms for dedicated problems in domains such as quantum chemistry or physical simulation, and provide the first step towards the dream of fault-tolerant quantum computing [126]. Among the different (but still similar) solutions currently developed by the companies mentioned above, IBM's approach (yielding the so-called *IBM Q devices*) stands out since it is the first one which already made devices publicly available (through a cloud access launched within their project *IBM Q* [73] in 2017). In the following, IBM Q devices serve as representatives, but devices of other companies have similar characteristics and features, and are subject to similar constraints to be satisfied.

IBM's infrastructure allows access to several devices that are all located in the Yorktown research lab, but are named after IBM office locations around the globe. The 5-qubit devices are named Yorktown and Tenerife (formerly called *IBM QX2* and *IBM QX4*), while the 16-qubit devices are named Rueschlikon (formerly called *IBM QX3* and, in a revised form *IBM QX5*) and Melbourne (currently providing only 14 of its qubits). Moreover, there exist 20-qubit devices named Austin and Tokyo, respectively, which are available for IBM's partners and members of the IBM Q network. In January 2019, IBM presented the *IBM Q System One* at CES [72]—the first 20-quit prototype towards a quantum computer for commercial use and to be operated outside of their labs.

Like the devices of Intel, Rigetti, and Google, IBM's quantum chips use superconducting qubits (based on Josephson junctions) that are connected with coplanar waveguide bus resonators [52, 70, 85, 135]. Quantum operations are conducted by applying microwave pulses to the qubits. By this, all these devices have the same (or at least similar) physical constraints that have to be satisfied when running quantum algorithms (i.e., *quantum circuits*) on them.

IBM Q devices support two types of quantum operations (i.e., quantum gates): The one-qubit gate $U(\theta, \phi, \lambda) = R_z(\phi)R_y(\theta)R_z(\lambda)$, which is composed of two rotations around the z-axis and one rotation around the y-axis (i.e., an Euler decomposition) allows to realize any unitary 2×2 matrix by specifying the parameters accordingly (up to global phase). For example, the Hadamard and T gate are realized as $H = U(\frac{pi}{2}, 0, \pi)$ and $T = U(0, 0, \frac{\pi}{4})$, respectively. To gain universality, IBM Q devices also support CNOT gates (controlled X gates) that allow coupling pairs of qubits.

In IBM's devices [74–78], a U gate is realized by three frame changes (FC) that are interleaved with two Gaussian derivative (GD) pulses. Note that the frame changes are equivalent to a virtual Z-gate in software and, thus, do not cause any operation to be conducted on the quantum hardware. Moreover, there exist cheaper realizations in case that some of the parameters of the U gate are set to specific values. More precisely, realizing a gate $U_2(\phi, \lambda) = U(\frac{\pi}{2}, \phi, \lambda)$ requires only two frame changes and one Gaussian derivative pulse, and a gate $U_1(\lambda) = U(0, 0, \lambda)$ requires only one frame change (thus, not requiring any operation on the actual hardware). Finally, CNOT gates are realized with one frame change, three Gaussian derivative pulses as well as two Gaussian flattop pulses (since cross-resonance interaction serves as basis for CNOT gates).

While one-qubit gates can be applied without limitations, the physical architecture of the respectively developed quantum computers—usually a linear or rectangular arrays of qubits—limits two-qubit gates to neighboring qubits that are connected by a superconducting bus resonator [85, 135]. In IBM's devices that use cross-resonance interaction as the basis for CNOT gates, the frequencies of the qubits also determine the direction of the gate (i.e., determining which qubit is the control and which is the target). The possible CNOT gates are captured by the so-called coupling maps [79], giving a very flexible description means to specify the *coupling-constraints* of a certain quantum device. Note that these constraints also do not undermine universality, but require to convert arbitrary circuits into compliant circuits with *local connectivity* only (which is covered in detail in Part IV of this book).

Definition 2.3 *Let* $V = \{Q_0, Q_1, \ldots, Q_{m-1}\}$ *be a set of m physical qubits available on the quantum device and let* $E \subseteq V \times V$ *defines which physical qubits can interact with each other. More precisely,* $(Q_i, Q_j) \in E$ *states that a CNOT gate with control qubit* Q_i *and target qubit* Q_j *can be applied, while* $(Q_i, Q_j) \notin E$ *states that a CNOT gate with control qubit* Q_i *and target qubit* Q_j *cannot be applied. Then, the directed graph* (V, E) *is the* coupling map *of the quantum device.*

Example 2.3 *Figure 2.2 shows the coupling maps for IBM's quantum devices. Physical qubits are visualized with nodes and a directed edge from physical qubit* Q_i *to physical qubit* Q_j *indicates that a CNOT with control qubit* Q_i *and target qubit* Q_j *can be applied.*

As mentioned above, the development of quantum devices is still in its infancy (despite great accomplishments recently achieved)—resulting in devices advocated to the NISQ era. Metrics that quantify the quality of a device's qubits are the accuracy with which operations can be applied to the qubits as well as time the qubit's state remain stable. One-qubit gates have error rates in the 10^{-3} range, while the error rates of two-qubit gates and measurement are in the 10^{-2} range. Typically, gates have execution times in the 10–100 ns range. The timespan of qubits remaining stable is quantified by the *energy relaxation time* T_1 (indicating how long it takes for a qubit to decay from the excited state $|1\rangle$ to the ground state $|0\rangle$) as well as the *dephasing time* T_2 (indicating how long the coupling between two qubits remains after setting them into superposition) [52]. Due to the currently achievable coherence time of qubits (T_1 and T_2 are in the 100 µs range) and gate fidelity, the number of gates to be executed as well as the number of sequentially applied gates are severely limited. Exact execution times of gates, gate error rates, and coherence times of IBM's devices are updated (calibrated) on a regular basis and can be found at [74–79].

Recently, researchers from Google claimed that they achieved quantum supremacy using their Sycamore quantum computer [8]. Although the claims in the paper might be a bit over-ambitious (as, e.g., discussed in recent work by competitors like IBM [123]), they show the significant progress made in the implementation of real quantum hardware and that physical realizations of NISQ devices that outperform conventional ones for practically relevant applications are within reach [30, 111, 126]. However, there exists even more potential for the quantum-computing paradigm by developing fault-tolerant devices that do not suffer from limitations like gate fidelity or coherence time of qubits. Such envisioned devices are able to conduct arbitrary deep computations with perfect accuracy—

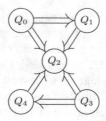

(a) IBM Q 5 Yorktown V1.2.0 (IBM QX2) [78]

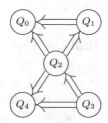

(b) IBM Q 5 Tenerife V1.3.0 (IBM QX4) [78]

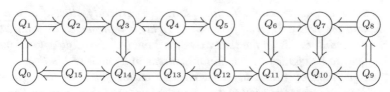

(c) IBM Q 16 Rueschlikon V1.0.0 (IBM QX3) [75]

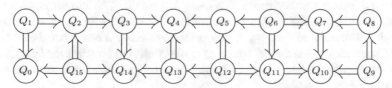

(d) IBM Q 16 Rueschlikon Version 1.1.0 (IBM QX5) [76]

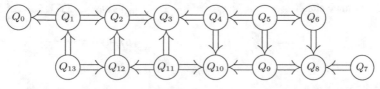

(e) IBM Q 16 Melbourne Version 1.1.0 [74]

Fig. 2.2 Coupling map of the IBM's quantum devices

allowing solving even more problems (e.g., integer factorization) in an efficient fashion. The key idea behind fault-tolerant quantum computing is to use error-correcting codes where several (up to hundreds) of physical qubits realize a single logical qubit of a computation [12, 56, 69, 125]. For example, this allows distilling a reasonably accurate quantum state out of several copies of imperfect states [50]. In this context, the universal Clifford+T gate library [19] got established.[5] This library is composed of the generators of the Clifford group, i.e., $CNOT$, H, and P, as well as the operation T (a rotation around the z-axis of the Bloch sphere by $\frac{\pi}{4}$ when neglecting global phase).

2.3   Reversible and Quantum Circuits

Quantum circuits [114] are used as proper description means for a finite sequence of "small" gates that cumulatively enact some unitary operator U and, given an initial state $|\varphi\rangle$ (which is usually the basis state $|0\ldots0\rangle$), produce a final state vector $|\varphi'\rangle = |U\varphi\rangle$. Hence, a quantum gate does not represent a physical entity (like in the conventional realm), rather an operation that is applied to a set of qubits.

Definition 2.4 *In quantum circuits, the qubits are vertically aligned in a circuit diagram, and the time axis (read from left to right) is represented by a horizontal line for each qubit. Boxes on the time axis of a qubit (or enclosing several qubits) indicate gates to be applied.[6] Note that measurement as reviewed in Sect. 2.1 also counts as quantum operation in this context. Control qubits of C-U gates are indicated by • and are connected to the controlled operations by a single line.*

Example 2.4 *Consider the quantum computation described in Example 2.2. Figure 2.3 shows the corresponding quantum circuit. The circuit contains two qubits, q_0 and q_1, which are both initialized with basis state $|0\rangle$. First, a Hadamard operation is applied to qubit q_0, which is represented by a box labeled H. Then, a CNOT operation is conducted, where q_0 is the control qubit (denoted by •) and q_1 is the target qubit (denoted by $\oplus$). Eventually, qubit q_0 is measured as indicated by the meter symbol.*

Fig. 2.3 Quantum circuit

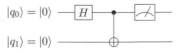

---

[5]Note that using the Clifford+T library any desired operation can be approximated up to a certain fixed error ϵ.
[6]Note that an X gate may also be denoted by $\oplus$.

Some quantum algorithms include large Boolean components (e.g., the modular exponentiation in Shor's algorithm [143]), where corresponding unitary matrices permute computational-basis vectors (hence, contain only 0s and 1s). Such circuits are composed of NOT (X) gates, "Controlled-NOT" (CNOT) gates, and the three-qubit CCNOT (Toffoli) gate defined by $|x\,y\,z\rangle \mapsto |x\,y\,xy \oplus z\rangle$, which exhibits nonlinear behavior due to the use of the Boolean *and* operation. These gates allow one to compute any reversible Boolean function—although this may require an extra 0-initialized qubit (called *ancilla*)—and form a so-called *reversible circuit*.

Definition 2.5 *A Boolean function* $f : \mathbb{B}^n \to \mathbb{B}^m$ *is* reversible, *if* $n = m$ *and the function is a bijection.*

This book uses the notation provided by the following definition for reversible circuits. Moreover, the terms quantum circuits and reversible circuits are used interchangeably whenever suitable, i.e., whenever a purely Boolean part of a quantum circuit is considered, it might also be denoted as reversible circuit.

Definition 2.6 *Let* $X = \{x_0, x_1, \ldots, x_{n-1}\}$ *be a set of* n *circuit lines. Then, a* reversible circuit *is a cascade* $G = g_0 g_1 g_2 \ldots g_{|G|-1}$ *of* $|G|$ *reversible gates* g_i. *A reversible gate (here:* multiple-controlled Toffoli gate) $g_i = TOF(C_i, t_i)$ *consists of a set* $C_i \subseteq \bigcup_{x_j \in X}\{x_j^+, x_j^-\}$ *of positive* (x_j^+) *and negative* (x_j^-) *control lines and a target line* $t_i \in X$ *with* $\{t_i^+, t_i^-\} \cap C_i = \emptyset$. *A circuit line cannot be used as positive and negative control at the same time, i.e.,* $\forall x_j \in X : x_j^+ \notin C_i \vee x_j^- \notin C_i$. *The value of the target line* t_i *is inverted iff the values of all positive control lines* $x_j^+ \in C_i$ *evaluate to 1 and all negative control lines* $x_j^- \in C_i$ *evaluate to 0. All lines other than the target line pass through the gate unchanged.*

In circuit diagrams, positive control lines, negative control lines, and the target line of a Toffoli gate are depicted using symbols ●, ○, and ⊕, respectively. Moreover, the Boolean elements 0 and 1 are used to denote values instead of the basis states $|0\rangle$ and $|1\rangle$ since reversible circuits can be treated entirely in the conventional domain.

Example 2.5 *Figure 2.4 shows a reversible circuit composed of three circuit lines and three Toffoli gates. Furthermore, the circuit is labeled with the values on the circuit lines for input* $x_0 x_1 x_2 = 001$. *The first gate* $g_0 = TOF(\{x_2^+\}, x_0)$ *inverts the value of the target line* x_0 *since the positive control line* x_2^+ *is assigned 1. For the same reason (control lines are accordingly assigned), the second gate* $g_1 = TOF(\{x_0^+, x_1^-\}, x_2)$ *inverts the value of the target line* x_2. *In contrast, the third gate* $g_2 = TOF(\{x_2^+\}, x_1)$ *does not invert the value of the target line* x_1, *because the positive control line* x_2^+ *evaluates to 0.*

Fig. 2.4 Reversible circuit

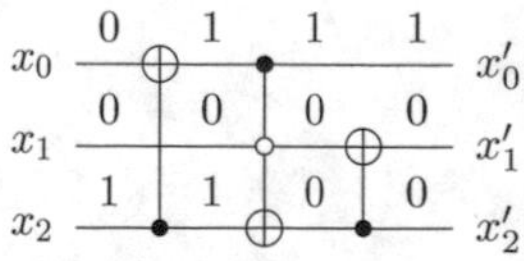

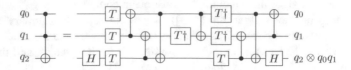

Fig. 2.5 Decomposition of a Toffoli gate [6]

Comparing the quality of quantum circuits representing a certain functionality requires cost metrics like the number of qubits or operations. While the number of required qubits shall be kept as small as possible for NISQ as well as for fault-tolerant devices (since qubits are a rather limited resource), other metrics have only importance for one of these two categories. For example, quantum algorithms relying on large Boolean components are expected to be executable on fault-tolerant quantum computers due to their large number of gates. Hence, occurring gates are usually decomposed into the Clifford+T library (see Sect. 2.2) for execution on quantum computers. Since Boolean reversible circuits generated by automated synthesis approaches usually contain multiple-controlled NOT gates with arbitrarily many positive and negative controls (cf. Part III), these *multiple-controlled Toffoli* gates are first decomposed into Toffoli, CNOT, and NOT (i.e., X) gates and, eventually, Toffoli gates are decomposed as shown in Fig. 2.5 using, e.g., methods proposed in [6, 11, 99, 106, 139, 180]. Since the *fault-tolerant* realization of the T operation is much more costly than for the other elements in the library, the overall number of T operations (i.e., *T-count*) or the number of subsequent T operations (i.e., *T-depth*) shall be kept as small as possible in the context of fault-tolerant computing. Thus, *T-count* and *T-depth* serve (among others) as cost metrics for quantum circuits, e.g., generated by automatic synthesis approaches (cf. Part III). For multiple-controlled gates, T-count and T-depth grows linearly with the number of control qubits, implying that their number shall be kept as small as possible in automated approaches.

Example 2.6 *As shown in Fig. 2.5, the T-count and T-depth of a Toffoli gate with two control lines are 7 and 3, respectively.*

In contrast, *T-count* and *T-depth* are not necessarily relevant for NISQ devices, since a $T = U_1(\frac{\pi}{4})$ is basically for free on the IBM devices (cf. Sect. 2.2). Instead, it is important to keep the overall depth of the circuit as well as the overall number of gates when designing circuits for NISQ devices, since the coherence time of qubits is limited and each gate introduces an error with a certain probability. For the latter, gate counts might also be weighted since the fidelity of two-qubit gates or measurement is much lower than the fidelity of one-qubit gates.

Chapter 3
Design Automation Methods for Conventional Systems

Abstract This chapter provides a rough overview on efficient data-structures and algorithms that are heavily used in design automation for conventional electronic systems. These algorithms are capable of efficiently solving exponential problems in many cases, and have inspired the algorithms presented in this book.

Keywords Design automation · Function representation · Search methods · Boolean satisfiability · Decision diagrams

This chapter briefly reviews methods utilized in design automation for the conventional realm (i.e., for electronic circuits and systems) that serve as basis for sophisticated hardware design tools as taken for granted today and constitute a main reason for the utilization and penetration of electronic devices into almost all parts of our daily life. This includes methods for compact representation of Boolean functionality, efficient search algorithms that allow pruning large parts of the search space, as well as powerful reasoning engines that can handle large problem instances.

3.1 Decision Diagrams

Boolean functions related to conventional electronic circuits and systems often contain a large number of inputs and outputs—making explicit function representations such as truth tables infeasible due to their exponential growth. Hence, efficient design automation tools have to rely on more compact description means that still allow to precisely represent the desired functionality (i.e., without losing information). One well-known techniques are *decision diagram*, where Boolean functions are represented as *directed acyclic graph* (DAG), whose structure is formed by recursively decomposing the function to be represented over its input variables. Eventually, compactness is gained by using shared nodes to represent equal (redundant) sub-functions.

© Springer Nature Switzerland AG 2020

A. Zulehner, R. Wille, *Introducing Design Automation for Quantum Computing*,
https://doi.org/10.1007/978-3-030-41753-6_3

Based on the initial description of *Binary Decision Diagrams* (BDDs [23]), different decomposition schemes have been heavily investigated in the 1990s—leading to different decision diagram types such as *Kronecker Function Decision Diagrams* (KFDDs, [42]), *Binary Moment Diagrams* (BMDs, [24]), or *Zero-suppressed Decision Diagrams* (ZDDs, [107]). They all are particularly suited for compactly representing certain types of Boolean functions—leading to efficient methods, e.g., in verification [24, 93] or synthesis [43]—even though their worst-case complexity (when no redundancies can be exploited) remains exponential. Corresponding implementations (usually denoted *DD-packages*) as provided by Fabio Somenzi's BDD-package CUDD [156], the Word-level-DD-package [66], or Donald Knuth's BDD-package [88] affect the development of design tools and methods until today.

In the following, BDDs serve as representatives for decision diagrams in general. To this end, consider multi-output Boolean functions $f \colon \mathbb{B}^n \to \mathbb{B}^m$, which are commonly given in terms of descriptions of their *primary outputs* $y_0, \ldots, y_{m-1}$. These m single-output Boolean functions $\mathbb{B}^n \to \mathbb{B}$ are commonly described in terms of Boolean algebra, i.e., as *Sums of Products* (SOP), *Products of Sums* (POS), or similar.

To gain a compact and canonic representation of the Boolean functions by means of BDDs, a fixed order $x_0 \prec x_1 \prec \ldots \prec x_{n-1}$ of the input variables is employed[1] and the single-output functions are represented as root nodes labeled by x_0 in the resulting DAG. The function f_v of a node v labeled by x_i is recursively defined (following the defined variable order) as

$$f_v = \left(x_i \wedge f_{high(v)} \right) \vee \left(\overline{x}_i \wedge f_{low(v)} \right),$$

where $f_{high(v)}$ and $f_{low(v)}$ denote the functions represented by the high and low child, respectively. This equation has a strong analogy to the *Shannon decomposition* of f (with respect to a primary input x_i) which is given as

$$f = \left(x_i \wedge f_{x_i=1} \right) \vee \left(\overline{x}_i \wedge f_{x_i=0} \right).$$

Here, $f_{x_i=1}$ and $f_{x_i=0}$ are the so-called *co-factors* of f which are obtained by setting the primary input x_i to 1 and 0, respectively. The analogy between the two equations, on the one hand, justifies the claim that the BDD-nodes represent the Shannon decomposition of f with respect to its primary inputs and, on the other hand, yields a blueprint for how to construct the BDD representation of a given function. Compactness is gained by using shared nodes to represent equal cofactors (sub-functions). Alternatively, as logical operations like AND, OR, etc., can be conducted directly and efficiently on BDDs (i.e., polynomial in the number of decision diagram nodes), the BDD representation of an SOP can also be constructed by first building the BDDs for the individual products and then using the BDD equivalent of the logical OR operation to "sum up" the products.

---

[1]The employed variable order may significantly affect the size of such a BDDs.

Fig. 3.1 Representations of a half adder

x_0	x_1	y_0	y_1
0	0	0	0
0	1	0	1
1	0	0	1
1	1	1	0

(a) Truth-table (b) BDD

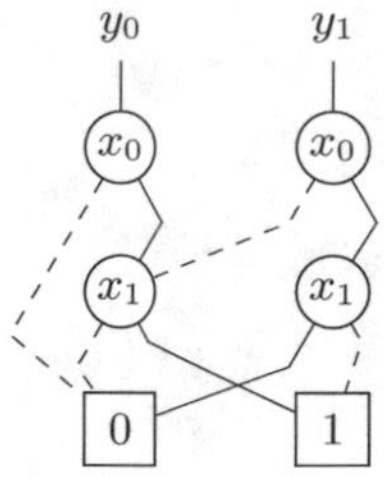

Example 3.1 *Consider the function of a* half adder *described by the multi-output Boolean function* $f : \mathbb{B}^2 \to \mathbb{B}^2$ *with output functions* $y_0(x_0, x_1) = x_0 \wedge x_1$ *(carry) and* $y_1(x_0, x_1) = x_0 \oplus x_1 = x_0\overline{x}_1 \vee \overline{x}_0 x_1$ *(sum). The corresponding truth table is shown in Fig. 3.1a. Figure 3.1b shows the BDD with variable order* $x_0 \succ x_1$ *for the half adder function. The output functions* y_0 *and* y_1 *are represented by means of root nodes. Consider the output* y_0 *of the half adder. The left node labeled* x_0 *represents the Shannon decomposition of* y_0 *with respect to* x_0. *Since the 0-cofactor (obtained by setting* $x_0 = 0$*) is* $f_{x_0=0} = 0$*, the dashed edge directly points to the 0-terminal. In contrast, the solid edge representing the 1-cofactor points to another node labeled* x_1 *since* $f_{x_0=1} = x_1$. *Applying this decomposition recursively and for* y_1 *results in the BDD shown in Fig. 3.1b. Compactness is gained by using shared nodes to represent equal cofactors (e.g., the left node labeled* x_1*).*

3.2 Heuristic Search Methods

Besides the existence of compact representation, e.g., for Boolean functions, efficient search methods for reasoning play a vital role in design automation methods for conventional logic. Since in many cases the search space is exponentially large, efficient search algorithms are required to avoid searching through "uninteresting" parts of the search space, while still guaranteeing an optimal (or at least a close to optimal) solution. This section revisits A* as representative for such search algorithms.

The A* algorithm [65] is an informed search algorithm where states in the search space of the considered problem are represented by nodes. Nodes that represent a solution are called *goal nodes* (multiple goal nodes may exist).[2] The main idea is to determine the cheapest path (i.e., the path with the lowest costs) from the root node to a goal node. Since the search space is typically exponential, sophisticated mechanisms are employed in order to consider as few paths as possible.

---

[2]Note that the path from the root node to a goal node may also determine the solution (instead of the goal node itself).

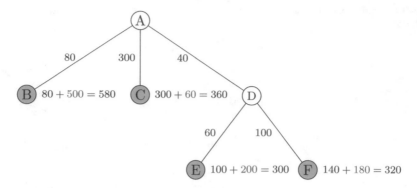

Fig. 3.2 A* search algorithm

All search algorithms are similar in the way that they start with a root node (representing the initial state) which is iteratively expanded towards a goal node (i.e., the desired solution). How to choose the node that shall be expanded next depends on the actual search algorithm. For example, A* search determines the costs of each leaf node of the already expanded search space. Then, the node x with the lowest costs $f(x) = g(x) + h(x)$ is chosen to be expanded next. Here, the first part ($g(x)$) describes the costs for reaching the current state from the initial one (i.e., the costs of the path from the root node to x). The second part describes the remaining costs to reach a goal node (i.e., the costs of the path from x to a goal node), which is estimated by a heuristic function $h(x)$. Since the node with the lowest costs is expanded, some parts of the search space (those that lead to expensive solutions) are never traversed.

Example 3.2 *Consider the tree shown in Fig. 3.2. This tree represents the part of the search space that has already been explored for a certain search problem. The nodes that are candidates to be expanded in the next iteration of the A* algorithm are highlighted in blue. For all these nodes, the costs $f(x) = g(x) + h(x)$ is determined as sum composed of the costs of the path from the root to x (i.e., the sum of the costs annotated at the respective edges) and the estimated costs of the path from x to a goal node (provided in red). Consider the node labeled E. This node has costs $f(E) = (40 + 60) + 200 = 300$. The other candidates labeled B, C, and F have costs $f(B) = 580$, $f(C) = 360$, and $f(F) = 320$, respectively. Since the node labeled E has the fewest expected cost, it is expanded next.*

Obviously, the heuristic costs should be as accurate as possible to expand as few nodes as possible. If $h(x)$ always provides the correct minimal remaining cost, only the nodes along the cheapest path from the root node to a goal node would be expanded. Since the minimal costs are usually not known (otherwise, the search problem would be trivial to solve), estimations are employed. However, to ensure an optimal solution, $h(x)$ has to be *admissible*, i.e., $h(x)$ must not overestimate the costs of the cheapest path from x to a goal node. This ensures that no goal node is expanded (which terminates the search algorithm) until all nodes that have the potential to lead to a cheaper solution are expanded.

Example 3.2 (Continued) *Consider again the node labeled E. If $h(x)$ is admissible, the true costs of each path from this node to a goal node is greater than or equal to 200.*

3.3 Efficient Reasoning Engines

Besides search methods as discussed above, many design automation methods rely on powerful reasoning engines that can also deal with large search spaces. A prominent representative are *Boolean satisfiability* (SAT) solvers that are capable of finding assignments to variables such that a Boolean expression evaluates to true, or proving that no such assignment is possible—becoming state of the art for many design automation problems like automatic test pattern generation [89, 142], logic synthesis [183], diagnosis [146], and verification [14, 29, 124].

Definition 3.1 *The* satisfiability problem *determines an assignment to the variables of a Boolean function $\Phi : \{0, 1\}^n \to \{0, 1\}$ such that Φ evaluates to 1 or proves that no such assignment exists. In an extended interpretation, an objective function $\mathcal{F}$ defined by $\mathcal{F}(x_1, \ldots, x_n) = \sum_{i=1}^{n} w_i \dot{x}_i$ with $w_1, \ldots, w_n \in \mathbb{Z}$ and $\dot{x} \in \{\overline{x}_i, x_i\}$ is additionally provided. In this case, an assignment is to be determined which does not only satisfy Φ but, at the same time, minimizes $\mathcal{F}$.*

Example 3.3 *Let $\Phi = (x_0 + x_1 + \overline{x}_2)(\overline{x}_0 + x_2)(\overline{x}_1 + x_2)$. Then, $x_0 = 1$, $x_1 = 0$, and $x_2 = 1$ is a satisfying assignment solving the SAT problem. Additionally, let $\mathcal{F} = x_0 + x_1 + x_2$. Then, $x_0 = 0, x_1 = 0$, and $x_2 = 0$ is a solution which does not only satisfies Φ but also minimizes $\mathcal{F}$.*

Despite SAT is a central $\mathcal{NP}$-complete problem [31], instances with hundreds of thousands of variables can often be solved by today's efficient reasoning engines (e.g., [39]). Key techniques for such efficient solvers are—besides the basic DPLL algorithm [37, 38]—intelligent decision heuristics [54], *conflict based learning* [96], and *Boolean Constraint Propagation* (BCP [113]). The general idea of the DPLL algorithm is to successively choose a value for an unassigned variable until all variables are assigned (yielding a solution for the problem instance), or until a conflict is caused that cannot be resolved using backtracking of previous decisions (the problem instance is proven to be *unsatisfiable*). Thereby, implications of assignments are determined using BCP.

In the past, very efficient reasoning engines for the satisfiability problem or its extension to *SAT Modulo Theories* (SMT [15]) problem have been proposed (e.g., [20, 39, 44, 172]). In order to utilize their power a symbolic formulation is required that completely describes the considered problem by means of a Boolean function.

Part II
Quantum-Circuit Simulation

Chapter 4
Overview

Abstract This chapter sets the context for the complementary quantum-circuit simulation approach that is based on *Decision Diagrams* (DDs) and presented in this part of the book. To this end, existing simulation tasks and strategies are presented— including a discussion of the resulting challenge and how they are handled in existing simulators. Moreover, this chapter provides a preview of the remaining chapters of this part of the book.

Keywords Quantum computer · Simulation · Quantum circuits · Quantum states · Qubits

Since physical realizations of quantum computers are limited in their availability, their number of qubits, their gate fidelity, and coherence time, quantum-circuit simulators running on conventional machines are required for many tasks. From a user's perspective, possible applications (or at least their prototypes) for quantum computers are usually first evaluated through simulators that serve as temporary substitute. Moreover, simulation can be adapted to circuit equivalence-checking and other functional verification tasks useful for circuit designers [118, 162, 167, 181]. Simulation also plays an important role for designers of quantum systems, e.g., to foster the development of error-correcting codes. Besides that, the urgent need of verifying quantum hardware might be conducted (at least some of the required verification tasks) by comparing runs on these machines to simulation outcome [1, 16, 34]. Ultimately, quantum-circuit simulation capabilities provide an estimate on *quantum supremacy* [8, 16, 123] as well as to identify classes of circuits where no quantum speedup is reachable (i.e., in case these circuits can be simulated efficiently on a conventional machine). In all these scenarios, simulators may give additional insights since, e.g., the precise amplitudes of a quantum state are explicitly determined (while they are not observable in a real quantum computer).

© Springer Nature Switzerland AG 2020

A. Zulehner, R. Wille, *Introducing Design Automation for Quantum Computing*,
https://doi.org/10.1007/978-3-030-41753-6_4

Overall, quantum-circuit simulation helps to

- evaluate and improve quantum computer architectures,
- evaluate and improve quantum algorithms,
- estimate the advantage that quantum computers attain over conventional computers,
- establish bounds on quantum speedups,
- study quantum error-correcting codes and fault-tolerant circuits, and
- adapt quantum circuits to physical architectures to moderate the impact of imperfections.

However, quantum-circuit simulation suffers from an exponential complexity since the Hilbert space modeling a quantum system's state grows exponentially with respect to the number of qubits. Hence, several simulation strategies and corresponding implementations that try to tackle this complexity have been developed in the last decade. This resulted in a broad variety of simulation approaches, which follow different formalisms of quantum physics, compute all or only some entries of the state vector, and that either in an exact or approximate fashion. In the following, *exact strong* simulation (a task to be explicitly defined later in Sect. 4.1) following the Schrödinger formalism (where states represented by unit-norm vectors are modified by multiplying unitary matrices representing gate operations) is considered. This simulation approach is popular because—once the exponentially large state vector and matrices can be represented—it scales linearly with circuit depth and lends itself to parallelization on GPUs or distributed architectures. However, the array-like representation of the state vector in current state-of-the-art simulators results in exponential memory requirements, which limits the number of qubits to be simulated to approximately 30 on a modern computer (and to 50 when considering supercomputers with petabytes of distributed memory) [145].

This part of the book presents a complementary simulation approach (still following Schrödinger-style exact strong simulation) that aims for overcoming this memory bottleneck. To this end, dedicated *Decision Diagrams* (DDs) are developed, which reduce the memory requirements by representing redundancies in the occurring vectors and matrices by means of shared nodes. This allows gaining significant improvements compared to straightforward realizations (relying on array-like representations) in many cases—often reducing the simulation time from several hours or days to seconds or minutes.

To set the context for the simulation approach presented in this part of the book, this chapter provides the basics of quantum-circuit simulation. More precisely, Sect. 4.1 defines the simulation tasks to be conducted and provides an overview of different simulation strategies including corresponding computational complexity as well as capabilities. Based on that, Sect. 4.2 discusses the resulting challenge and reviews how state-of-the-art simulators following Schrödinger-style exact strong simulation cope with it. Finally, Sect. 4.3 provides the general idea of the presented complementary simulation approach and summarizes the gained improvements compared to the state of the art.

4.1 Simulation Tasks and Strategies

In general, a quantum computer that executes a given circuit and measures all (or some) qubits produces a non-deterministic bitstring on every run—sampled from a distribution defined by the final quantum state.[1] This behavior is sketched in Fig. 4.1a and can be captured by *exact weak simulation*, which produces bitstrings sampled from the distribution defined by the quantum circuit. Recognizing that quantum computers are susceptible to errors might also allow the simulation to produce inaccurate results. To this end, *approximate weak simulation* allows for a small statistical distance between the desired and the actual distribution.

Even though the amplitudes of the final quantum state are not exposed by the quantum computer (hence, not a required output of *weak simulation*), it is common to compute them explicitly and deterministically using *strong simulation* so as to

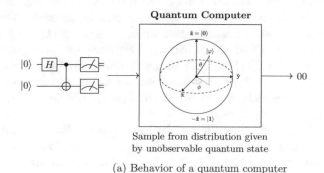

(a) Behavior of a quantum computer

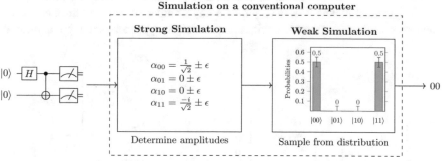

(b) Simulation of a quantum computer (using weak and strong simulation)

Fig. 4.1 Functional behavior of a quantum computer and its simulation

---

[1]As the final state is unambiguously determined by the quantum circuit, so is the outcome distribution.

facilitate probabilistic sampling of bitstrings.[2] Figure 4.1b illustrates the connection
between weak and strong simulation, compared to the behavior of an actual quantum
computer discussed before and sketched in Fig. 4.1a. Moreover, strong simulation
may be used for verification of quantum computers. For example, approaches developed in [1, 16, 34] evaluate bitstring samples against exact amplitudes/probabilities
computed deterministically on a non-quantum computer.

Techniques for strong and weak quantum-circuit simulation often draw upon
core paradigms of quantum mechanics developed by Schrödinger, Feynman, and
Heisenberg [49, 114, 138]. More precisely:

The Schrödinger formalism represents quantum states by their wave-functions,
while unitary operators modify these states. Simulation based on this formalism
results in a full set of 2^n complex amplitudes representing the wave-function of
an n-qubit quantum system as illustrated in Fig. 4.2a. Thereby, quantum gates are
represented by small unitary matrices that can be extended to full size ($2^n \times 2^n$)
by tensor products with identity matrices. To simulate such a gate, one multiplies
its extended matrix by the state vector (cf. Sect. 2.1), perhaps, using an algorithm optimized for such tensor-product structure, without explicitly storing the
full matrix (using $O(1)$ additional memory). Software tools that implement this
approach include LIQ$Ui|\rangle$ [168], QX [86], qHiPSTER [145], QuEST [84], and many
others. Moreover, Schrödinger-style simulators are included in quantum computing
frameworks like ProjectQ [157], Microsoft QDK [101], IBM Qiskit [33], and Rigetti
Forest [130].

Schrödinger-style simulation is popular because it is relatively straightforward,
scales linearly with circuit depth, and lends itself to parallelization on GPUs and
distributed architectures. Moreover, it easily provides full information for (error-
free) circuits of up to approximately 30 qubits on a modern computer (relying
on an array-like representation of vectors and matrices). Having the probabilities
of all outcomes makes it easy to produce any number of samples through biased
random selection. The problem with this simulation approach is that computing 2^n

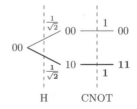

(a) Following the Schrödinger formalism   (b) Following the Feynman formalism

Fig. 4.2 Strong simulation motivated by core formalisms of quantum mechanics

---

[2]Techniques to perform weak simulation directly are rare and often adapt techniques from strong
simulation [22, 133, 159].

probabilities and even saving them becomes impossible (on current supercomputers) for $n > 50$ qubits when using straightforward algorithms, while a quantum computer does not actually make them available [145].

The Feynman formalism computes amplitudes/probabilities of individual outcomes without computing intermediate quantum states. Instead of mapping the initial state to a superposition after a quantum gate is applied, only one basis state from such a superposition is considered at a time. Applying this idea recursively entails a depth-first search over a tree of possibilities. Each quantum gate may create multiple path branches (in practice, diagonal and permutation gates do not trigger branching), and each path ends when all gates have been processed. By computing the product of relevant matrix elements on each path, one computes the path's contribution to one amplitude. These contributions must be integrated/summed over all possible evolution paths. For circuits with relatively few gates, careful path pruning helps scaling simulation up to 50 qubits on a laptop [141]. Besides computing multiple amplitudes in parallel, Feynman-path summation itself is easy to parallelize, since individual processes may follow individual paths. Aaronson and Chen [1] proposed a related polynomial-space simulation of n-qubit circuits of depth d. Using only $O(n \log d)$ space, their algorithm runs in $O(n \cdot (2d)^{n+1})$ time.[3] However, Feynman-path summation methods make poor use of the available memory. Therefore, approaches that hybridized Feynman-path summation with Schrödinger-style simulation to obtain a flexible space-time trade-off have been developed [1, 25, 94, 122]. These approaches are particularly suited for simulating quantum circuits proposed for conducting quantum supremacy experiments (proposed by researchers from Google in [8, 16]) due to their limited depth.

The Heisenberg formalism represents a quantum state by a group of unitary operators (from a limited space) that stabilize (leave unchanged) this state. This group may be finite and represented by its generators. Unitary operators acting on states are represented by their actions on those generators. In a given Hilbert space, this formalism is effective for only a finite number of states and operators and, thus, is somewhat limited in applicability. This formalism was developed into an algorithmic approach by Gottesman, Knill, and Aaronson for simulating circuits composed of gates from the Clifford group (i.e., circuits that are composed of H, P, and $CNOT$ gates only) in polynomial time [2, 7]. However, these circuits do not allow for universal quantum computation.

Besides the straightforward realizations of Schrödinger-style and Feynman-style simulation, there have also been developed approaches following these formalisms while their complexity adapts to the given circuits. Here, approaches based on *redundancy removal* [163], Matrix Product States (MPS [165]) [166], or *tensor network contraction* [17, 95] have shown significant improvements in many cases.

---

[3]Run-time is further reduced to $2^{O(d\sqrt{n})}$ for grid-based quantum circuits that loosely match architectures pursued by IBM, Google, Intel, etc.

In the following, this book focuses on Schrödinger-style exact strong simulation since it is straightforward and scales linearly with circuit depth—allowing simulating any quantum circuit composed whose exponentially large state vector and matrices can be represented. Mathematically spoken, Schrödinger-style simulation boils down to multiplying several matrices to the vector representing the initial state, eventually determining the amplitudes of the final state explicitly and in an exact fashion (i.e., exact strong simulation).

4.2 Resulting Challenge and State of the Art

This section discusses the resulting challenge of Schrödinger-style exact strong quantum-circuit simulation and reviews how related work handles them. To this end, recall Example 2.2 on Page 13. Here, the state of the two-qubit system to be simulated is represented by a 4-dimensional vector. The final state of the system is determined by multiplying 4×4-dimensional matrices representing the respective gate operations to the vector. Since the first operation is an H gate acting on a single qubit, the Kronecker product with I_2 (assuming the identity operation for the other qubit) is formed to gain a matrix of proper size. Eventually, a vector results that holds the amplitudes for all four basis states—allowing to draw samples from this distribution if desired (i.e., in case weak simulation shall be conducted).

Accordingly, most state-of-the-art simulators use straightforward representations like 1- and 2-dimensional arrays for vectors and matrices and, hence, are termed *array-based approaches* in the following. However, the arrays required to store the state vector and the unitary gate matrices grow exponentially with respect to the number of qubits—forming the main challenge.[4] Hence, simulation results for quantum systems with only up to 34 qubits (requiring 270 GB of memory) were reported on a single machine.

In order to simulate quantum systems composed of more qubits, solutions exploiting massive hardware power (supercomputers composed of thousands of nodes and more than a petabyte of distributed memory) are applied. Even then, quantum systems with less than 50 qubits are today's practical limit [145]. Overall, the current state-of-the-art Schrödinger-style approaches cope with this challenge as follows:

- *LIQUi|⟩* [168]: Microsoft's tool suite for quantum computing with an integrated simulator that relies on a straightforward representation and, thus, also can simulate systems with up to approximately 30 qubits only, when used on a Desktop machine with 32 GB RAM (still requires substantial run-time of up to several days).

---

[4]Note that the exponentially large matrix is usually not represented explicitly by state-of-the-art simulators due to its tensor-product structure.

- *qHiPSTER* [145]: A quantum high-performance software testing environment developed in Intel's Parallel Computing Lab. Here, parallel algorithms are utilized which are executed on 1000 compute nodes with 32 TB RAM distributed across these nodes. Even with this massive hardware power, quantum systems of rather limited size (not more than 40 qubits) can be simulated.
- *QX* [86]: A high-performance array-based quantum computer simulation platform developed in the QuTech Computer Engineering Lab at Delft University. The simulator tries to parallelize the application of quantum gates to improve the performance. The authors state that *QX* allows for simulation of 34 fully entangled qubits on a single node using 270 GB of memory.
- *ProjectQ* [157]: ProjectQ is a software framework for quantum computing that started at the ETH Zurich. The contained high-performance array-based simulator allows simulating up to approximately 30 qubits on a Desktop machine. ProjectQ additionally contains an emulator [64], which can determine, e.g., the result for Shor's algorithm significantly faster than the simulator by employing conventional shortcuts (e.g., arithmetic components are computed conventionally instead of using a quantum circuit). Here, simulation results are provided for systems with up to 36 qubits—accomplished with massive hardware power, i.e., a supercomputer similar to the one used for *qHiPSTER*.

Overall, array-based approaches remain limited due to the exponential growth with respect to the number of qubits (independent of the considered circuit). To remove this limitation, complementary approaches are required.

4.3   A Complementary Simulation Approach

This part of the book presents a complementary Schrödinger-style simulation approach that utilizes *Decision Diagrams* (DDs) to overcome the exponential growth of the occurring vectors and matrices—significantly outperforming the current state-of-the-art approaches outlined above in many practically relevant cases. Note that several types of decision diagrams have been developed for the quantum realm, e.g., *X-decomposition Quantum Decision Diagrams* (XQDDs, [167]), *Quantum Decision Diagrams* (QDDs, [3]), *Quantum Information Decision Diagrams* (QuIDDs, [164]), or *Quantum Multiple-valued Decision Diagrams* (QMDDs, [104, 115]). Moreover, some of them have even been considered for Schrödinger-style simulation [68, 132, 163]—following the key idea of representing occurring state vectors and unitary matrices more compactly by exploiting redundancies. One representative is *QuIDDPro* [163], which allows, e.g., to simulate Grover's algorithm significantly faster than with array-based solutions by exploiting certain redundancies in the occurring state vectors and unitary matrices. However, none of these so-called *graph-based* approaches got established thus far due to their limited applicability (i.e., they provide improvements in rather few cases).

This changed with the DD-based simulation approach presented in this book. More precisely, using a decomposition scheme that allows exploiting even more redundancies than previous simulators allows for significant speedups for a broader variety of quantum circuits to be simulated. While there obviously remain cases where the array-based approaches perform better (whenever no redundancies can be exploited), the presented simulator got established in the community as complementary approach, which is witnessed by a *Google Faculty Research Award* as well as an official integration into IBM's SDK Qiskit and Atos' Quantum Learning Machine (Atos QLM [10]).

Chapter 5 (based on [121, 198]) describes the basic ideas and required algorithms of the presented DD-based simulator, and shows that for many cases, the simulation time can be reduced from several days to just a few seconds or minutes. This initial implementation of a DD-based simulator did not only lead to a significant improvement compared to the current state of the art, but has also received significant acknowledgement by the community—triggering further optimizations as done for array-based Schrödinger-style simulators for more than a decade. In fact, Chap. 6 shows that the complexity of multiplications depends on the size (i.e., the number of nodes) of the respective operands in DD-based simulation. Together with the fact that the DDs for the usually considered gate matrices are linear in size (with respect to the number of qubits), this implies that it might be beneficial to combine gate operations before applying them to the state vector. Chapter 6 (based on [200]) describes strategies for combining operations that allow improving the initial version of the presented DD-based simulator significantly—up to several orders of magnitude when exploiting application-specific knowledge.

Moreover, it remains obvious that an efficient implementation of the underlying DD-package is paramount to actually turn the theoretical foundations of the presented DD-based approach discussed in Chaps. 5 and 6 into the significant speedups reported there. While many concepts of DD-packages developed for the conventional domain can be directly utilized, handling the occurring complex numbers requires new and non-trivial strategies since numerical errors might "shadow" redundancies (i.e., they cannot be detected), which potentially lead to an exponential blow-up of the DD. Therefore, Chap. 7 (based on [186]) presents methods to efficiently handle complex numbers in DDs—leading to a fully fletched DD-package for the quantum realm.

Last but not least, Chap. 8 (based on [120, 187]) provides an alternative approach to handle occurring complex numbers by providing an algebraic representation related to the Clifford+T library. This does not only allow evaluating the trade-off between accuracy and precision in DDs for the quantum realm present thus far, but also to simulate Clifford+T circuits in an exact fashion.

Chapter 5
Decision Diagram-Based Simulation

Abstract Decision diagrams for the quantum domain aim for providing a more compact representation of the exponentially large matrices and vectors by exploiting occurring redundancies through shared nodes. This is particularly beneficial for quantum-circuit simulation, which heavily relies on matrix–vector multiplication and, hence, on efficient means for that. This chapter provides the general idea of such a DD-based simulation approach. Results show that this significantly outperforms current state-of-the-art simulators in many practically relevant cases.

Keywords Qubits · Simulation · Quantum state · State vector · Vector · Matrix · Vector–matrix multiplication · Decision diagram · Compact representation

Schrödinger-style *Decision Diagram-based* (DD-based) simulation tries to represent the occurring state vectors and unitary matrices more compactly (compared to array-based implementations) by exploiting redundancies. First implementations have already shown significant improvements compared to straightforward Schrödinger-style simulators (relying on an array-based representation) for certain cases [68, 132, 163], but did not get established due to their limited applicability. This chapter (based on [121, 198]) presents a new type of decision diagram for the simulation of quantum circuits (inspired by Niemann et al. [115]) that utilizes a decomposition scheme that is more natural to the occurring matrices and vectors—allowing to exploit even more redundancies and, thus, represent them in a more efficient fashion. Since the natural decomposition scheme also allows realizing dedicated manipulation algorithms efficiently, this leads to a new simulation method, which clearly outperforms the current state of the art in many cases.

In the following, the considered complementary simulation approach is described. To this end, Sect. 5.1 describes the utilized decision diagrams to represent occurring vectors and matrices more compactly, while Sect. 5.2 describes DD-based algorithms to eventually conduct simulation (i.e., how to generate DDs for gate matrices, how to multiply matrices to vectors, and how to measure qubits).

© Springer Nature Switzerland AG 2020
A. Zulehner, R. Wille, *Introducing Design Automation for Quantum Computing*,
https://doi.org/10.1007/978-3-030-41753-6_5

Section 5.3 provides a discussion of the complexity of the representations and the operations compared to array-based approaches as well as to previous DD-based ones. Eventually, Sect. 5.4 provides an evaluation of the presented approach and a comparison to the state of the art, using frequently considered quantum algorithms as benchmarks.

5.1 Representations for Quantum-Circuit Simulation

This section discusses the utilized representation for state vectors and unitary matrices required for quantum-circuit simulation. To this end, a compact representation for state vectors is presented, which, afterwards, is extended by a second dimension—leading to a compact representation for quantum operations.

5.1.1 Representation of State Vectors

As discussed in Sect. 4.1, the Schrödinger formalism represents quantum states by their wave-functions that are composed of 2^n complex amplitudes—an exponential representation. However, taking a closer look at state vectors unveils that they are frequently composed of redundant entries, which provide the basis for a more compact representation.

Example 5.1 *Consider a quantum system with $n = 3$ qubits situated in a state given by the following vector:*

$$\varphi = \left[0, 0, \frac{1}{2}, 0, \frac{1}{2}, 0, -\frac{1}{\sqrt{2}}, 0\right]^T .$$

Although of exponential size ($2^3 = 8$ entries), this vector only assumes three different values, namely 0, $\frac{1}{2}$, and $-\frac{1}{\sqrt{2}}$.

Such redundancies can be exploited for a more compact representation. To this end, decision diagram techniques (similar to those from the conventional realm and discussed in Sect. 3.1) are employed. More precisely, similar to decomposing a function into sub-functions, a given state vector with its complex entries is decomposed into sub-vectors. To this end, consider a quantum system with qubits $q_0, q_1, \ldots q_{n-1}$, whereby w.l.o.g. q_0 represents the most significant qubit. Then, the first 2^{n-1} entries of the corresponding state vector represent the amplitudes for the basis states with q_0 set to $|0\rangle$; the other entries represent the amplitudes for states with q_0 set to $|1\rangle$. This decomposition is represented in a decision diagram structure by a node labeled q_0 and two successors leading to nodes representing the sub-vectors. The sub-vectors are recursively decomposed further until vectors of size 1 (i.e., a complex number) result. This eventually represents the amplitude α_i for the complete basis state and is given by a terminal node. During these decompositions, equivalent sub-vectors are represented by the same node—allowing for sharing and, hence, a reduction of the memory complexity. An example illustrates the idea.

Example 5.2 *Consider again the quantum state from Example 5.1. Applying the decompositions described above yields a decision diagram as shown in Fig. 5.1a. The left (right) outgoing edge of each node labeled q_i points to a node representing the sub-vector with all amplitudes for the basis states with q_i set to $|0\rangle$ ($|1\rangle$). Following a path from the root to the terminal yields the respective entry. For example, following the path highlighted bold in Fig. 5.1a provides the amplitude for the basis state with $q_0 = |1\rangle$ (right edge), $q_1 = |1\rangle$ (right edge), and $q_2 = |0\rangle$ (left edge), i.e., $-\frac{1}{\sqrt{2}}$ which is exactly the amplitude for basis state $|110\rangle$ (seventh entry in the vector from Example 5.1). Since some sub-vectors are equal (e.g., $\left[\frac{1}{2}, 0\right]^T$ represented by the left node labeled q_2), sharing is possible.*

However, even more sharing is possible since sub-vectors often differ in a common factor only (e.g., the state vector from Example 5.1 has sub-vectors $\left[\frac{1}{2}, 0\right]^T$ and $\left[-\frac{1}{\sqrt{2}}, 0\right]^T$, which differ by the factor $-\sqrt{2}$ only). This is additionally exploited in the considered representation by denoting common factors of amplitudes as weights attached to the edges of the decision diagram. Then, the value of an amplitude for a basis state is determined by following the path from the root to the terminal, and additionally multiplying the weights of the edges along this path. Again, an example illustrates the idea.

Example 5.3 *Consider again the quantum state from Example 5.1 and the corresponding decision diagram shown in Fig. 5.1a. As can be seen, the sub-vectors*

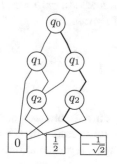

(a) Without edge weights

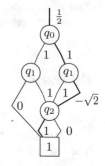

(b) With edge weights

Fig. 5.1 DD-based representation of state vectors

represented by the nodes labeled q_2 (i.e., $\left[\frac{1}{2}, 0\right]^T$ and $\left[-\frac{1}{\sqrt{2}}, 0\right]^T$) differ in a common factor only.

In the decision diagram shown in Fig. 5.1b, both sub-trees are merged. This is possible since the corresponding value of the amplitudes is now determined not by the terminals, but the weights on the respective paths. As an example, consider again the path highlighted bold representing the amplitude for the basis state $|110\rangle$. Since this path includes the weights $\frac{1}{2}$, 1, $-\sqrt{2}$, and 1, an amplitude of $\frac{1}{2} \cdot 1 \cdot (-\sqrt{2}) \cdot 1 = -\frac{1}{\sqrt{2}}$ results.

Note that, of course, various possibilities exist to factorize an amplitude. Hence, a normalization is applied which assumes the left edge to inherit a weight of 1. More precisely, the weights w_l and w_r of the left and right edge are both divided by w_l and this common factor is propagated upwards to the parents of the node. If $w_l = 0$, the node is normalized by propagating w_r upwards to the parents of the node.[1]

The resulting representation discussed above leads to the following definition.

Definition 5.1 *The resulting decision diagram for representing a 2^n-dimensional state vector is a directed acyclic graph with one terminal node labeled 1 that has no successors and represents a 1-dimensional vector with the element 1. All other nodes are labeled q_i, $0 \leq i < n$ (representing a partition over qubit q_i) and have two successors. Additionally, there is an edge pointing to the root node of the decision diagram. This edge is called* root edge. *Each edge of the graph has attached a complex number as weight. An entry of the state vector is then determined by the product of all edge weights along the path from the root towards the terminal. Without loss of generality, the nodes of the decision diagram are ordered by the significance of their label, i.e., the successors of a node labeled q_i are labeled with a less significant qubit q_j. The decision diagram is reduced, i.e., nodes where both outgoing edges point to the same successor and have attached the same weight (i.e., both sub-vectors are equal) are removed. Finally, the nodes are normalized, which means that all edges-weights are divided by the first non-zero weight. The common factor is propagated upwards in the decision diagram.*

5.1.2 Representation of Matrices

As discussed in Sect. 2.1, quantum operations acting on n qubits are described by $2^n \times 2^n$-dimensional unitary matrices. Similar to state vectors, these matrices often include redundancies, which allow for a more compact representation. To this

---

[1] Applying a fixed normalization scheme, a representation which is even canonic (w.r.t. qubit order) results. However, since canonicity is irrelevant for the purpose of simulation, this issue is not discussed in detail here.

end, the utilized decomposition scheme for state vectors is extended by a second dimension—yielding a decomposition scheme for matrices.

The entries of a unitary matrix $U = [u_{i,j}]$ indicate how basis state $|i\rangle$ is mapped to a basis state $|j\rangle$ (i.e., how much it contributes to the respective amplitude). Considering again a quantum system with qubits $q_0, q_1, \dots q_{n-1}$, whereby w.l.o.g. q_0 represents the most significant qubit, the unitary matrix U acting on all n qubits is decomposed into four sub-matrices of dimension $2^{n-1} \times 2^{n-1}$: All entries in the left upper sub-matrix (right lower sub-matrix) provide the values describing the mapping from basis states $|i\rangle$ to $|j\rangle$ (i.e., $|i\rangle \rightarrow |j\rangle$) with both assuming $q_0 = |0\rangle$ ($q_0 = |1\rangle$). All entries in the right upper sub-matrix (left lower sub-matrix) provide the values describing the mapping from basis states $|i\rangle$ with $q_0 = |1\rangle$ to $|j\rangle$ with $q_0 = |0\rangle$ ($q_0 = |0\rangle$ to $q_0 = |1\rangle$). This decomposition is represented in a decision diagram structure by a node labeled q_0 and four successors leading to nodes representing the sub-matrices. The sub-matrices are recursively decomposed further until a 1×1 matrix (i.e., a complex number) results. This eventually represents the value $u_{i,j}$ for the corresponding mapping. During these decompositions, equivalent sub-matrices are again represented by the shared nodes and a corresponding normalization scheme (as applied for the representation of state vectors) is employed. Note that for a simpler graphical notation, zero stubs indicate zero matrices (i.e., matrices that contain zeros only) and edge weights that are equal to one are omitted. Again, an example illustrates the idea.

Example 5.4 *Consider the matrices of* H, I_2, *and* $U = H \otimes I_2$ *(cf. Example 2.2). Figure 5.2 shows the corresponding decision diagram representations. Following the path highlighted bold in Fig. 5.2c defines the entry* $u_{0,2}$*: a mapping from* $|0\rangle$ *to* $|1\rangle$ *for* q_0 *(third edge from the left) and from* $|0\rangle$ *to* $|0\rangle$ *for* q_1 *(first edge). Consequently, the path describes the entry for a mapping from* $|00\rangle$ *to* $|10\rangle$*. Multiplying all factors on the path (including the* root *edge) yields* $\frac{1}{\sqrt{2}} \cdot 1 \cdot 1 = \frac{1}{\sqrt{2}}$*, which is the value of* $u_{0,2}$*.*

The concepts described above lead to the definition of a decision diagram representing a unitary matrix as follows.

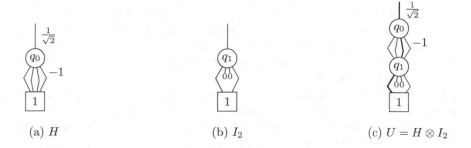

(a) H (b) I_2 (c) $U = H \otimes I_2$

Fig. 5.2 DD-based representation of matrices

Definition 5.2 *The resulting decision diagram for representing a* $2^n \times 2^n$-
*dimensional unitary matrix is a directed acyclic graph with one terminal node
labeled 1 that has no successors and represents a* 1×1 *matrix with the element
1. All other nodes are labeled* q_i, $0 \leq i < n$ *(representing a partition over qubit
q_i) and have four successors. Additionally, there is an edge pointing to the root
node of the decision diagram. This edge is called* root edge. *Each edge of the
graph has attached a complex number as weight. An entry of the unitary matrix is
then determined by the product of all edge weights along the path from the root
towards the terminal. Without loss of generality, the nodes of the decision diagram
are ordered by the significance of their label, i.e., the successors of a node labeled
q_i are labeled with a less significant qubit q_j. The decision diagram is reduced,
i.e., nodes where all outgoing edges point to the same successor and have attached
the same weight (i.e., all four sub-matrices are equal) are removed. Finally, the
nodes are normalized, which means that all edges-weights are divided by the
first non-zero weight. The common factor is propagated upwards in the decision
diagram.*

5.2   Conducting Quantum-Circuit Simulation

With the availability of a compact representation for state vectors and unitary
matrices, it is left to provide corresponding methods for building gate matrices,
for applying them to the current state of the considered quantum system, as well as
for measuring qubits. These required operations can be efficiently implemented on
the utilized DDs since the applied decomposition scheme is natural to vectors and
matrices.

5.2.1   Constructing Decision Diagrams Representing Gate
Matrices

This book considers circuits composed of gates that implement controlled one-qubit
operations (C-U gates with $t = 1$) as input for the presented DD-based simulator
(cf. Sect. 2.1), which is common for simulators. This does not only allow simulating
circuits composed of elementary gates if desired (e.g., by limiting the number of
controls to 1), but also to avoid decomposition of high-level C-U gates (e.g., Toffoli
gates) if this not required for the considered simulation purpose—increasing the
performance of the simulator since fewer gates are applied. Consequently, a gate
$g_i = CU(\mathbf{U_i}, C_i, t_i)$ in a circuit composed of n qubits is fully specified by a
2×2 unitary matrix $\mathbf{U_i}$, a target qubit $t_i \in \{q_0, \ldots, q_{n-1}\}$, as well as a set of
positive and negative control qubits $C_i \subset \{q_0^+, \ldots, q_{n-1}^+\} \cup \{q_0^-, \ldots, q_{n-1}^-\}$ (with
$\{q_t^+, q_t^-\} \cap C_i = \emptyset$).

Example 5.5 *Consider a quantum circuit that is composed of three qubits q_0, q_1, and q_2 and gates $g_0 = CU(\mathbf{H}, \emptyset, q_2)$ and $g_1 = CU(\mathbf{X}, \{q_2^+\}, q_1)$ as shown in Fig. 5.3a. First, a Hadamard operation is applied to qubit q_2, followed by a CNOT (CX) gate with control qubit q_2 and target qubit q_1. Hence, the circuit entangles qubits q_2 and q_1.*

The DD for a quantum gate is built variable by variable (qubit by qubit) in a bottom-up fashion from the terminal to the root node. To this end, the variable order $q_0 \succ q_1 \succ \ldots \succ q_{n-1}$ is assumed from the root node towards the terminal node. The notation $\mathbf{M}_{\{q_k,\ldots,q_{n-1}\}}$ (describing a $2^{n-k} \times 2^{n-k}$-dimensional matrix) indicates the set of variables that have been processed so far. Moreover, for the sake of an easier reference, those edges of a DD-node labeled q_c that correspond to a mapping of states $|i\rangle \rightarrow |j\rangle$ where both assume $q_c = |0\rangle$ or $q_c = |1\rangle$, i.e., the first and the fourth edge, are termed as *diagonal*. The remaining edges are termed *off-diagonal*.

Though it is possible to construct the DD for the gate $CU(\mathbf{U}, C, q_t)$ in a single run, for a better understanding two DDs are constructed representing the cases that the gate is active (all control qubits are in their activating state) or inactive (at least one control qubit is not). By adding these DDs, the actual DD for the gate results.

Case "gate is active", i.e., the 2×2 unitary matrix $\mathbf{U}$ is performed on qubit q_t if, and only if, all controls are in their activating state. All other qubits preserve their original state. Consequently, the DD for the active case contains all (non-zero) paths of the final DD for which all decision variables (qubits) except for the target have an activating assignment.

In order to have a valid starting point, the algorithm begins at the terminal level with an edge pointing to the terminal node with weight 1, i.e., $\mathbf{M}_\emptyset = [1]_{1 \times 1}$. Afterwards, the qubits are processed in a bottom-up fashion. If the current qubit q_c

- is neither a control nor the target, i.e., $q_c \neq q_t, q_c^+ \notin C, q_c^- \notin C$, the gate is active regardless of the qubit's state. Consequently, at the matrix level the result is $\mathbf{I}_2 \otimes \mathbf{M}_{\{q_{c+1},\ldots,q_{n-1}\}}$ which corresponds to a DD-node labeled q_c where all diagonal edges point to the existing DD and all remaining edges are 0-edges.
- is a positive (negative) control qubit, i.e., $q_c^+ \in C$ ($q_c^- \in C$), the gate is only active for $q_c = |1\rangle$ ($q_c = |0\rangle$). Consequently, the result is a node labeled q_c with only 0-edges except the fourth (first) one, which points to the existing DD.
- is the target, i.e., $q_c = q_t$, the transition described by $\mathbf{U}$ is performed. Consequently, the result is $\mathbf{U} \otimes \mathbf{M}_{\{q_{c+1},\ldots,q_{n-1}\}}$, i.e., a node labeled q_t with all edges pointing to the existing DD with the corresponding edge weight taken from the unitary matrix $\mathbf{U}$ (if a weight is zero, the corresponding edge is a 0-edge directly pointing to the terminal).

During this construction, the DD is normalized as described in Sect. 5.1.2.

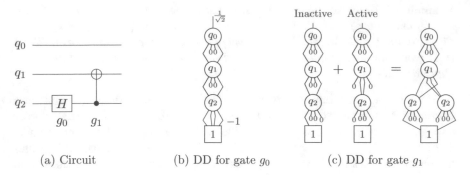

(a) Circuit   (b) DD for gate g_0 (c) DD for gate g_1

Fig. 5.3 DD-based representation for quantum gates

Example 5.5 (Continued) *Consider the DD in Fig. 5.3b, which represents the first gate of the quantum circuit g_0 discussed above. As this gate does not have any controls, it is always active and, thus, it suffices to build the DD representing the active part. Building the active part starts with an edge to the terminal node with weight 1. As the bottom-most qubit is already the target qubit, all edges of the q_2-node point directly to this terminal with the appropriate weight of the Hadamard transformation matrix $\mathbf{H} = \frac{1}{\sqrt{2}}\left(\begin{smallmatrix} 1 & 1 \\ 1 & -1 \end{smallmatrix}\right)$. Note that normalization will propagate the common factor $\frac{1}{\sqrt{2}}$ of this matrix to the root edge. The remaining qubits are neither control nor target. Thus, nodes representing an identity mapping of these qubits are inserted—forming $I_2 \otimes I_2 \otimes H$.*

The DD for the inactive case is constructed similarly.

Case "gate is inactive",   i.e., the identity transition is performed on qubit q_t since at least one control is not in its activating state. All qubits preserve their original state, i.e., none but diagonal edges are populated at all. Consequently, the DD for the inactive case contains all (non-zero) paths of the final DD for which at least one decision variable (qubit) does not have an activating assignment.
However, when constructing the DD in a bottom-up fashion, the hypothesis is always used that all controls above the current qubit are in their activating states and at least one control below is not. To make sure that this hypothesis gives the correct result even for the bottom-most control (for which no inactive control may exist below), the algorithm starts at the terminal level with an edge pointing to the terminal node with weight 0, i.e., $\mathbf{M}_\emptyset = [0]_{1 \times 1}$. This ensures that all edges corresponding to the activating value of this bottom-most control are 0-edges. The remaining qubits are processed as follows. If the current qubit q_c

- is neither a control nor the target, i.e., $q_c \neq q_t, q_c^+ \notin C, q_c^- \notin C$, the gate is inactive regardless of the qubit's state. Consequently, at the matrix level the result is $I_2 \otimes M_{\{q_{c+1},...,q_{n-1}\}}$ which corresponds to a DD-node labeled q_c

where all diagonal edges point to the existing DD and all remaining edges are 0-edges.

- is a positive (negative) control, i.e., $q_c^+ \in C$ ($q_c^- \in C$), the gate is definitely inactive for $q_c = |0\rangle$ ($q_c = |1\rangle$). Consequently, the result is a node with all diagonal edges pointing to the k-fold ($k = n-c-1$) tensor product $I_2^{\otimes k} = I_{2^k}$ (nothing happens to all k qubits below the current one) except from the fourth (first) edge. The latter handles the case that the qubit is in its activating state and is pointing to the existing DD $M_{\{q_{c+1},\dots,q_{n-1}\}}$.[2] All off-diagonal edges are 0-edges.

- is the target, i.e., $q_c = q_t$, the identity transformation is performed on the target. Consequently, the result is $I_2 \otimes M_{\{q_{c+1},\dots,q_{n-1}\}}$ like in the unconnected case.

Example 5.5 *The DD for the circuit's second gate g_1 is shown in Fig. 5.3c. For the inactive part, the algorithm starts with a 0-edge. For the positive control q_2^+, node is constructed which uses this 0-edge as fourth edge, while first edge represents the identity $I_2^{\otimes 0} = [1]_{1 \times 1}$, i.e., it points to the terminal node with weight 1. As q_2 is the only control, nodes representing an identity mapping for the remaining qubits are added.*

For the active part, the algorithm starts with an edge to the terminal node, which becomes the fourth edge of the node labeled q_2, as the activating state of q_2 is $|1\rangle$. For the target qubit q_1 with the transition matrix $X = \begin{pmatrix} 0 & 1 \\ 1 & 0 \end{pmatrix}$, a q_1-node is added. For this node, both off-diagonal edges point to the q_2-node constructed before (with weight 1 as the corresponding entry in X is 1) and both diagonal edges are 0-edges (as the corresponding entry in X is 0). Last, but not least, for the unconnected qubit q_0 a node representing its identity mapping is added. Finally, adding the DDs for the inactive and active part yields the actual DD for the CNOT gate.

Overall, the resulting DDs for the active as well as the inactive part of the gate are linear in the number of variables—regardless of the complexity of the gate under consideration. Both DDs can be constructed in parallel while iterating through the variables in a bottom-up fashion and describe disjoint parts of the gate matrix, while they are padded with zeros outside of that particular part. Consequently, their sum can be computed in linear time and is linear in size. In fact, there are only trivial additions where at least one of the addends is a 0-matrix and, as already recognized in [104], the addition could be saved entirely, such that the whole construction could be performed in a single pass from the terminal to the root node with no backtracking or recursion. Either way, DD-based representations for single gates are computed very efficiently.

Moreover, the algorithms for constructing the gate matrices show that forming the Kronecker product of two matrices $A \otimes B$ is inherently trivial on the considered DD-based representation. One only has to replace the terminal node of A with the

---

[2]If there is no further control below the current qubit, the gate inactivity is ensured by choosing a 0-edge as the initial DD.

root node of B (and multiply the weight of B's root edge to the weight of A's root edge), as shown in Fig. 5.2.

5.2.2   Multiplying Matrices and Vectors

Like matrices and vectors themselves, matrix–vector multiplication (required to determine the resulting state vector when applying a gate) can also be decomposed with respect to the most significant qubit, leading to

$$U \cdot \varphi = \begin{bmatrix} U_{00} & U_{01} \\ U_{10} & U_{11} \end{bmatrix} \cdot \begin{bmatrix} \varphi_0 \\ \varphi_1 \end{bmatrix} = \begin{bmatrix} U_{00} \cdot \varphi_0 \\ U_{10} \cdot \varphi_0 \end{bmatrix} + \begin{bmatrix} U_{01} \cdot \varphi_1 \\ U_{11} \cdot \varphi_1 \end{bmatrix}.$$

This means, that the four sub-products $U_{00} \cdot \varphi_0$, $U_{01} \cdot \varphi_1$, $U_{10} \cdot \varphi_0$, and $U_{11} \cdot \varphi_1$ are recursively determined first.[3] As shown in Fig. 5.4, these sub-products are then combined with a decision diagram node to two intermediate state vectors. Finally, these intermediate state vectors are added. This addition is recursively decomposed in a similar fashion, namely

$$\varphi + \psi = \begin{bmatrix} \varphi_0 \\ \varphi_1 \end{bmatrix} + \begin{bmatrix} \psi_0 \\ \psi_1 \end{bmatrix} = \begin{bmatrix} \varphi_0 + \psi_0 \\ \varphi_1 + \psi_1 \end{bmatrix}.$$

The recursively determined sub-sums $\varphi_0 + \psi_0$ and $\varphi_1 + \psi_1$ are composed by a decision diagram node as shown in Fig. 5.5.

Moreover, all these decompositions into sub-products and sub-sums are inherently given by taking the corresponding edges of the currently considered DD-node. Since redundant sub-products and sub-sums are computed only once (by utilizing caching as described in detail in Chap. 7), the computational complexity remains bounded by the number of nodes of the original representation of the matrix and the vector.

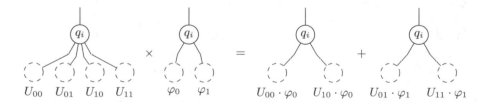

Fig. 5.4 Multiplication of a unitary matrix and a state vector

---

[3]The decompositions of multiplication and addition are recursively applied until 1×1 matrices or 1-dimensional vectors result. Since those eventually represent just complex numbers, their multiplication and/or addition is straightforward.

Fig. 5.5 Addition of state vectors

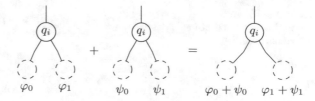

5.2.3 Measuring Qubits

Measurement is also conducted directly on the decision diagram structure. Since measuring the considered quantum system (i.e., drawing a sample according to the represented probability distribution) is equal to measuring all qubit sequentially (cf. Sect. 2.1), this task is conducted by traversing the DD from the root node to the terminal while choosing the left (right) edge of the node labeled q_i according to the probabilities for measuring $q_i = |0\rangle$ ($q_i = |1\rangle$). This requires computing the respective probabilities at first hand.

Consider qubit q_0 of the state vector (which is represented by the root node of the corresponding DD). Then, the left (right) successor of the root node represents the sub-vector containing the amplitudes of all states with $q_0 = |0\rangle$ ($q_0 = |1\rangle$), i.e., states that are of form $|0q_1q_2\ldots\rangle$ ($|1q_1q_2\ldots\rangle$). The probability for collapsing qubit q_0 to one of the two basis states is determined by summing over the probabilities of all (would-be) full-measurement outcomes for the remaining qubits (cf. Sect. 2.1), i.e.,

$$P(q_0 \rightarrow |0\rangle) = \sum_{x \in 0\{0,1\}^{n-1}} \alpha_x \alpha_x^*$$

$$P(q_0 \rightarrow |1\rangle) = \sum_{x \in 1\{0,1\}^{n-1}} \alpha_x \alpha_x^*.$$

Determining these probabilities requires summing up the probabilities (i.e., the squared magnitude of the respectively considered amplitude) of all paths through the left (right) edge of the root node. A recursive decomposition of this sum is given by

$$\sum_{x \in 0\{0,1\}^{n-1}} \alpha_x \alpha_x^* = \sum_{x \in 00\{0,1\}^{n-2}} \alpha_x \alpha_x^* + \sum_{x \in 01\{0,1\}^{n-2}} \alpha_x \alpha_x^*.$$

Moreover, the amplitudes α_x of the 2^n basis states are represented as product of $n + 1$ edge weights, i.e., $\alpha_x = \prod_{i=0}^{n} w_{x,i}$, in the utilized DDs. Since $(\alpha \cdot \beta)^* = \alpha^* \cdot \beta^*$ holds for all complex numbers $\alpha, \beta \in \mathbb{C}$, $\alpha_x \alpha_x^*$ is determined on the decision diagram by

$$\alpha_x \alpha_x^* = \left(\prod_{i=0}^{n} w_{x,i}\right)^* = \prod_{i=0}^{n} (w_{x,i})^*.$$

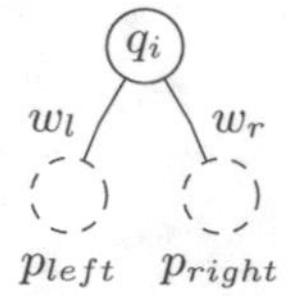

Fig. 5.6 Upstream
probability p of a decision
diagram node

$$p = p_{left} \cdot w_l w_l^* + p_{right} \cdot w_r w_r^*$$

These two recursive decompositions allow computing the so-called *upstream probability* (i.e., the summed probability of all paths from the currently considered node to the terminal node) of all DD-nodes in a depth-first traversal (i.e., in linear time with respect to the number of DD-nodes). More precisely, the upstream probability p of a node is determined as sum of the upstream probabilities of its successors (sub-vectors) p_{left} and p_{right}—weighted by the squared magnitude of the respectively attached edge weight as shown in Fig. 5.6.

The weighted upstream probabilities of the left and the right successor of the root node (after multiplying them with the squared magnitude of the root edge weight) determine the probabilities $P(q_0 \rightarrow |0\rangle)$ and $P(q_0 \rightarrow |1\rangle)$—allowing to sample a basis state for q_0. If $q_0 = |0\rangle$ ($q_0 = |1\rangle$) is chosen, the left (right) edge is taken to reach the qubit (node) to be measured next by using the same methodology.[4] This procedure is repeated until all qubits are measured. Eventually, a new DD representing the measured state is built in a bottom-up fashion—a trivial task since the measured state is a basis state and, hence, contains a single node for each qubit.

Example 5.6 *The decision diagram shown in Fig. 5.1b represents the quantum state φ discussed in Example 5.1. The upstream probability of the node labeled q_2 is $1^2 + 0^2 = 1$. Based on that, the upstream probabilities of the two nodes labeled q_1 are determined. These are $0^2 \cdot 1 + 1^2 \cdot 1 = 1$ for the left node and $1^1 \cdot 1 + \left|-\sqrt{2}\right|^2 \cdot 1 = 3$ for the right node. From these nodes, the probabilities for collapsing q_0 to basis state $|0\rangle$ or $|1\rangle$ are given by*

$$P(q_0 \rightarrow |0\rangle) = \left(\frac{1}{2}\right)^2 \cdot 1^2 \cdot 1 \quad = \frac{1}{4}$$

$$P(q_0 \rightarrow |1\rangle) = \left(\frac{1}{2}\right)^2 \cdot 1^2 \cdot 3 \quad = \frac{3}{4}$$

Hence, $|0\rangle$ and $|1\rangle$ are chosen as measurement outcome for q_0 with a probability of 0.25 and 0.75, respectively. Assuming that the measurement outcome for q_0 is $|1\rangle$, the left edge is taken to determine the measurement outcome for q_1. The determined upstream probabilities of the successors (in conjunction with the corresponding edge weights and the upstream probability of the considered node) give probabilities

---

[4]Note that it is not required to compute the upstream probabilities for each qubit measured since they are all already computed by the initial depth-first traversal of the DD.

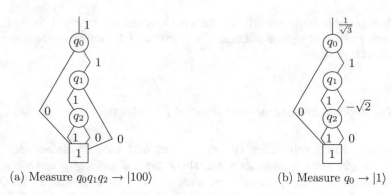

(a) Measure $q_0q_1q_2 \to |100\rangle$ (b) Measure $q_0 \to |1\rangle$

Fig. 5.7 Measurement of the state shown in Fig. 5.1b

$P(q_1 \to |0\rangle | q_0 \to |1\rangle) = (1^2 \cdot 1)\frac{1}{3} = \frac{1}{3}$ and $P(q_1 \to |1\rangle | q_0 \to |1\rangle) = ((-\sqrt{2})^2 \cdot 1)\frac{1}{3} = \frac{2}{3}$. Similarly, the measurement outcome for q_2 is determined after sampling a value for q_1. Assuming that $q_0q_1q_2 \to |100\rangle$ has been measured, the DD shown in Fig. 5.7a results.

Measuring a single qubit is also conducted directly on the DD. If the most significant qubit is measured, one follows the strategy described above, but stops after determining the value for q_0. If another qubit q_i is measured, the so-called *downstream probabilities* (i.e., the summed probability of all paths from the root node to the currently considered node) need to be determined for all nodes labeled q_i before conducting the measurement. This is necessary to weight the upstream probabilities of the nodes labeled q_i before summing them up. Depending on the sampled basis state for qubit q_i, measurement is conducted by applying a one-qubit gate with one of the non-unitary transformation matrices $\mathbf{M_0} = \frac{1}{\sqrt{P(q_i \to |0\rangle)}}\left(\begin{smallmatrix} 1 & 0 \\ 0 & 0 \end{smallmatrix}\right)$ or $\mathbf{M_1} = \frac{1}{\sqrt{P(q_i \to |1\rangle)}}\left(\begin{smallmatrix} 0 & 0 \\ 0 & 1 \end{smallmatrix}\right)$. These matrices set all the amplitudes of all basis states with $q_i = |0\rangle$ ($q_i = |1\rangle$) to 0 and adjust the root edge weight such that the state vector maintains a norm of 1.

Example 5.6 (Continued) *Assume only qubit q_0 shall be measured and that basis state $|1\rangle$ is sampled (which has probability $P(q_0 \to |1\rangle) = \frac{3}{4}$). Figure 5.7a shows the resulting decision diagram after applying the one-qubit gate $M_1 = \frac{2}{\sqrt{3}}\left(\begin{smallmatrix} 0 & 0 \\ 0 & 1 \end{smallmatrix}\right)$— resulting in the DD shown in Fig. 5.7b.*

5.3 Discussion

This section discusses the complexity of the presented DD-based simulation approach with respect to existing array-based and graph-based Schrödinger-style solutions. This allows getting an intuition why DD-based simulation as presented in this chapter significantly outperforms them in many cases. Thereby, a discussion on the representation of vectors and matrices as well as a discussion on the respective operations is provided.

5.3.1 Representation of Vectors and Matrices

While array-based approaches always have to deal with 2^n-dimensional state vectors, graph-based approaches often allow for a significantly more compact representation in many practically relevant cases. This is similar to BDDs in conventional design, which have an exponential size for most Boolean functions (especially random ones), but allow rather compact representation for many functions of interest. As already mentioned in Sect. 5.1.1, introducing edge weights and a normalization scheme allows exploiting more redundancies than previous graph-based solutions—leading to an even more compact representation.

Nevertheless, in the worst case (i.e., if no redundancies in the state vector can be exploited), a full binary tree with $|v| = 1 + \sum_{i=0}^{n-1} 2^i = 2^n$ nodes results. Furthermore, $2 \cdot (2^n - 1) + 1 = 2^{n+1} - 1$ complex edge weights have to be stored—approximately twice as many complex numbers than used in array-based solutions and when using, e.g., QuIDDs to represent the state vector. That is, in the absolute worst case, the utilized representation is twice as large as array-based and graph-based solutions known thus far. This additional overhead, however, allows exploiting many more redundancies and, hence, eventually gaining representations that are more compact. This is confirmed by a conducted evaluation, which clearly shows that the peak node count of the presented DD-based approach is significantly below that exponential upper bound in many practically relevant cases.

In general, the worst-case memory complexity for $2^n \times 2^n$ matrices is analogous to the memory complexity for state vectors—a full quad-tree has $|v| = 1 + \sum_{i=0}^{n-1} 4^i = 1 + \frac{4^n-1}{3}$ nodes. However, multiple-controlled one-qubit gates (as defined in Sect. 5.2), which are composed of a single target qubit and an arbitrary number of control qubits, require a linear (with respect to the number of qubits) number of nodes only. In most array-based simulators, the size of these matrices is also not an issue since only the 2×2 transformation matrix as well as the controls have to be stored—the exponentially large matrix does not have to be constructed explicitly. Hence, the crucial parameter that mainly determines the required memory consumption is the size of the state vector's representation.

5.3.2 Conducting Operations

Conducting operations is also less complex for graph-based solutions compared to array-based ones, which always suffer from an exponential complexity. However, there are also differences between the individual graph-based solutions.

As discussed in Sect. 5.2, the Kronecker product of two matrices can be easily determined on the utilized DDs by exchanging the terminal of one matrix with the root node of the other matrix. Consequently, forming the Kronecker product has complexity of $O(|v|)$. In contrast, this is more complex when using, e.g., QuIDDs [164], where it is required—among others—to multiply all terminal values of the two matrices with each other.

In addition, matrix vector multiplication can be performed significantly faster for graph-based solutions. The complexity of a matrix multiplication has an upper bound defined by the product of the number of nodes needed to represent the state vector and the number of nodes needed to represent the matrix. Since the number of nodes of the considered matrices grows linearly with the number of qubits, the overall complexity is $O(n \cdot |v|)$.

Measuring all qubits of a state vector has also complexity that is dominated by the size of the DD. First, the DD is traversed in a depth-first (breadth-first) fashion to determine the upstream (downstream) probabilities for all nodes. Then, each qubit is measured in $O(1)$, starting at the top of the decision diagram—resulting in an overall complexity of $O(|v| + n)$.

Overall, this clearly shows that graph-based solutions offer more efficient representation and manipulation of state vectors in many cases. Although graph-based approaches suffer from overhead caused by the decision diagram structure (and additional complex numbers), these approaches often outperform array-based solutions through a much more compact representation by exploiting redundancies. Since the presented DD-based approach exploits even more redundancies than, e.g., QuIDDPro, a significant performance improvement compared to this representative is observed. This is confirmed by a conducted evaluation summarized in the next section.

5.4 Evaluation

This section evaluates the scalability of the presented DD-based approach and compares it to state-of-the-art Schrödinger-style simulators (array-based ones as well as previous graph-based ones). To this end, the simulator has been implemented in C++.[5] Details on the implementation of the underlying DD-package are provided later in Chap. 7. The state of the art is represented by the publicly available implementations of the recently proposed array-based simulators *LIQUi|⟩* [168], *QX* [86], and the simulator of *ProjectQ* [157],[6] as well as the graph-based simulator *QuIDDPro* [163]. All simulations have been conducted on a regular Desktop

---

[5]The implementation is publicly available at http://iic.jku.at/eda/research/quantum_simulation.

[6]Note that ProjectQ also provides an emulator, where high-level operations are applied directly. However, in order to allow for a fair evaluation, the simulators are compared to each other and not a simulator to an emulator.

computer, i.e., a 64-bit machine with 4 cores (8 threads) running at a clock frequency of 3.8 GHz and 32 GB of memory running Linux 4.4.[7] The timeout was set to 5 h. Moreover, the best results published for other simulators (taken from the respective papers) are considered.

As benchmarks serve well-known quantum algorithms considered by previous publications. More precisely, quantum systems generating entangled states, conducting *Quantum Fourier Transformation* (QFT; cf. [114]), executing Grover's Algorithm for database search [59], and executing Shor's Factorization Algorithm [143] (using the realization proposed by Beauregard [13] that requires $2n + 3$ qubits to factor an n-bit integer) have been considered. Note that, for all benchmarks except QFT, the initial assignments of the inputs are fixed. For QFT, one of the basis states is randomly chosen as initial input assignment.

Table 5.1 summarizes the results. The columns *#Qubits* and *#Ops* list the number of involved qubits and the number of quantum operations to be conducted, respectively. Moreover, benchmark-specific parameters that influence the performance of the presented DD-based simulator are listed for Shor's algorithm.[8]

The remaining columns list the simulation time (i.e., the entire run-time from initialization to termination) and peak memory when using LIQ$Ui|\rangle$, QX, ProjectQ, QuIDDPro, as well as the presented DD-based approach. For the graph-based simulators (i.e., QuIDDPro as well as the presented one), the table includes a column listing the peak node count during simulation, which gives a more accurate measurement of scalability than the actual memory consumption. Besides that, Table 5.1 also provides a comparison to other state-of-the-art simulation approaches. That is, whenever results from them are provided in the literature, the respectively best result for a considered quantum algorithm is summarized in the bottom of that part.

Note that some restrictions apply for certain state-of-the-art simulators: In fact, the publicly available version of LIQ$Ui|\rangle$ allows simulating circuits composed of at most 23 qubits only. QX is able to simulate up to 29 qubits on the used machine—trying to allocate 30 or more qubits failed (due to limited memory). Similarly, ProjectQ allows simulating up to 31 qubits. However, QX has the additional drawback that it does not allow simulating Beauregard's realization of Shor's algorithm for integer factorization, because of missing features in the circuit description language (since QX is still in its infancy). All these cases are accordingly marked by *n.a.* (not applicable) in Table 5.1. Furthermore, the current release of QuIDDPro (version 3.8) also contains an improved simulator called *QuIDDProLite* (to be activated with the command line option *-cs*) that runs on average four times faster than QuIDDPro. However, this improved version can only simulate stand-

---

[7]The presented DD-based approach as well as QuIDDPro use a single core, while the simulators LIQ$Ui|\rangle$, QX, and the simulator of ProjectQ use multiple threads.

[8]This includes the number to be factored N as well as another number a co-prime to N, for which the multiplicative order r is determined by the algorithm to compute the prime factors of N. The maximal size of the state vector is influenced by r and the bitwidth of N [166].

Table 5.1 Evaluation

Computation	#Qubits	#Ops	LIQ $Ui\rangle$ [168]		QX [86]		Project Q [157]		QuIDDPro [163]			Presented DD-based approach		
			Time [s]	Mem [MB]	Time [s]	Mem [MB]	Time [s]	Mem [MB]	Time [s]	Mem [MB]	#Nodes	Time [s]	Mem [MB]	#Nodes
Entanglement	22	22	3.53	193.33	0.42	200.47	1.08	152.41	0.04	14.93	45	<0.01	96.55	43
	23	23	4.09	248.25	0.80	396.94	0.49	248.01	0.04	14.92	47	<0.01	96.55	45
	24	24	n.a.	–	1.61	790.23	0.64	444.69	0.04	14.93	49	<0.01	96.70	47
	29	29	n.a.	–	63.02	25,169.79	9.12	12,634.67	0.04	14.92	59	<0.01	96.64	57
	30	30	n.a.	–	MO	–	17.76	25,217.66	0.04	14.93	61	<0.01	96.65	59
	31	31	n.a.	–	MO	–	MO	–	0.04	14.93	63	<0.01	96.70	61
	100	100	n.a.	–	MO	–	MO	–	0.14	15.98	201	0.02	98.06	199
	● Reported for QX [86]: max. 34 qubits using less than 270 GB of memory													
QFT	21	231	6.46	192.83	3.63	102.75	0.66	100.75	10,208.06	2511.26	4194 303	0.02	97.36	21
	22	253	11.38	191.00	8.06	201.01	1.06	150.30	MO	–	–	0.02	97.36	22
	23	276	22.43	312.41	15.06	397.50	1.55	250.18	MO	–	–	0.02	97.72	23
	24	300	n.a.	–	31.66	790.77	3.27	445.04	MO	–	–	0.02	97.66	24
	29	435	n.a.	–	1270.03	25,170.21	109.19	12,638.54	MO	–	–	0.04	98.3	29
	30	465	n.a.	–	MO	–	234.27	25,217.39	MO	–	–	0.04	98.34	30
	31	496	n.a.	–	MO	–	MO	–	MO	–	–	0.04	98.99	31
	64	2080	n.a.	–	MO	–	MO	–	MO	–	–	0.85	104.22	64
	● Reported for qHiPSTER (Intel, cf. [145]): max. 40 qubits on a supercomputer (in approx. 1000s)													
	● Reported for Quantum emulator from [64]: max. 36 qubits on a supercomputer (in approx. 10 s)													

(continued)

Table 5.1 (continued)

Computation	#Qubits	#Ops	LIQ $Ui\rangle$ [168] Time [s]	Mem [MB]	QX [86] Time [s]	Mem [MB]	Project Q [157] Time [s]	Mem [MB]	QuIDDPro [163] Time [s]	Mem [MB]	#Nodes	Presented DD-based approach Time [s]	Mem [MB]	#Nodes
Grover	18	26,083	770.26	193.42	583.45	144.41	16.37	58.12	23.49	17.24	44	1.56	134.80	84
	20	57,941	8494.59	198.22	6394.54	382.51	77.95	77.17	85.86	19.05	39	3.89	134.54	94
	21	86,038	TO	–	TO	–	229.45	101.27	168.07	20.60	n.a.	6.12	134.60	99
	25	40,9626	n.a.	–	TO	–	15,765.00	843.85	2598.08	40.55	n.a.	53.67	108.94	119
	26	602,395	n.a.	–	TO	–	TO	–	5076.64	52.95	n.a.	85.72	109.20	124
	27	884,764	n.a.	–	TO	–	TO	–	TO	–	–	131.39	109.82	129
	30	2780,431	n.a.	–	MO	–	TO	–	TO	–	–	478.52	106.41	144
Shor ($N = 21, a = 2$)	15	12,635	298.59	62.72	n.a.	–	9.30	51.43	16,236.14	365.22	65 535	0.53	98.36	60
($N = 39, a = 2$)	17	38,847	343.83	128.81	n.a.	–	19.25	55.10	TO	–	–	1.01	99.39	94
($N = 497, a = 8$)	21	92,700	7888.00	147.03	n.a.	–	187.14	115.74	TO	–	–	30.96	131.30	531
($N = 1007, a = 529$)	23	134,490	TO	–	n.a.	–	947.72	283.88	TO	–	–	23.28	119.03	253
($N = 1561, a = 8$)	25	188,757	n.a.	–	n.a.	–	4827.00	860.96	TO	–	–	92.85	131.40	705
($N = 3679, a = 16$)	27	258,060	n.a.	–	n.a.	–	TO	–	TO	–	–	367.42	132.23	2 233
($N = 8193, a = 1024$)	31	451,563	n.a.	–	n.a.	–	MO	–	TO	–	–	104.65	131.42	349
($N = 29,843, a = 16$)	33	581,272	n.a.	–	n.a.	–	MO	–	TO	–	–	8135.62	137.01	23 506

● Reported for LIQ $Ui\rangle$ (Microsoft, cf. [168]): max. 31 qubits (in more than 30 days)

Time denotes the actual run-time and not the CPU seconds (which would be even higher for [86, 157, 168] since they use parallelization to speed-up simulation). The timeout was set to 5 h alone quantum circuits and, hence, is not applicable for simulating Beauregard's implementation of Shor's algorithm (which also requires non-quantum control structures). Consequently, the listed run-time for Shor's algorithm is obtained using QuIDDPro, while the run-time of the other benchmarks is obtained using QuIDDProLite (i.e., always the best result of both QuIDDPro versions is reported). Since QuIDDProLite does not offer the capability to dump the peak node count, the listed number of nodes are obtained by QuIDDPro for simulation. In the cases where this simulation does not succeed within the given timeout of 5 h, the corresponding entry is labeled by *n.a.* (not applicable).

As can be seen, the simulation of quantum systems generating entangled states and conducting QFT (when assuming a basis state as input) shows a linear behavior on the presented DD-based simulator. While this allows for a rather unlimited scalability using the simulation approach presented in this chapter, Microsoft's simulator LIQ$Ui|\rangle$, QX, as well as ProjectQ show exponential behavior. Even massive hardware power such as employed by Intel's simulator *qHiPSTER* [145] (running on a machine with 1000 nodes and 32 terabytes of memory) or the quantum emulator of [64] (running on a similar machine) manages to conduct QFT for a maximum of 40 qubits only (and additionally requires hundreds of seconds, while the approach presented in this chapter terminates in a fraction of a second).

The graph-based simulator QuIDDPro [163] is capable of efficiently conducting entanglements (similar to the presented DD-based approach, only a linear amount of nodes is required). However, an exponential behavior can be observed when conducting QFT. This is caused by the fact that the state vector contains exponentially many different entries, which cannot be handled efficiently by QuIDDPro. In contrast, the presented solution can exploit further redundancies here (namely sub-vectors which are multiples of each other as discussed in Sect. 5.1)—resulting in a linear number of decision diagram nodes for QFT.

The simulation of Grover's Algorithm and Shor's Algorithm constitutes a more challenging task. But even here, the presented representation remains rather compact. For example, in case of simulating Shor's Algorithm with 33 qubits (choosing $N = 29843$ and $a = 16$), only slightly more than 23,500 nodes are required. In fact, the significantly larger number of operations is more challenging here. Nevertheless, the presented DD-based approach still manages to simulate both algorithms significantly more efficient and for more qubits than previous simulators. While, e.g., Microsoft's simulator LIQ$Ui|\rangle$ is capable of conducting Shor's Algorithm for at most 31 qubits in more than 30 days (on a similar machine; cf. [168]), the DD-based simulation approach presented in this chapter completes this task within less than 2 min (choosing the same values for N and a).

Also the previously proposed graph-based solution QuIDDPro cannot reach this efficiency. While QuIDDPro allows for a rather compact representation when simulating Grover's Algorithm, it requires a substantial amount of run-time. This is caused by the fact that the required operations cannot be conducted as efficiently as with the presented DD-based simulator (since its decomposition scheme does not allow exploiting as many redundancies in the state vectors; cf. Sect. 5.3). Hence, the presented DD-based approach clearly outperforms QuIDDPro (which works particularly well for Grover's Algorithm and has mainly been evaluated on that in the literature). Shor's Algorithm unveils similar limitations for QuIDDPro than for array-based solutions (in fact, LIQ$Ui|\rangle$ is performing even better than QuIDDPro in this case). Again, the DD-based approach presented in this chapter can handle all these cases much faster and for more qubits than before.

Overall, the presented DD-based simulation approach clearly outperforms previous Schrödinger-style simulators in terms of run-time as well as in terms of memory when redundancies can be exploited, since the significantly smaller representation allows outweighing the constant overhead caused by the DD structure. For these cases, the presented approach is also able to simulate quantum circuits with more qubits than other Schrödinger-style simulators. Moreover, this has been achieved on a single core of a regular Desktop machine, i.e., without massive hardware power or the utilization of supercomputers.

Chapter 6
Combining Operations in DD-Based Simulation

Abstract This chapter aims to push the capabilities of the proposed DD-based simulator further—despite it has already shown significant improvements compared to the state of the art. To this end, strategies are developed that combine several quantum gates before applying them to the state vector. This technique works particularly well for DD-based simulation and allows reducing simulation time by several factors, or even by orders of magnitude when application-specific knowledge is exploited.

Keywords DD-based simulation · Quantum circuits · Multiplication · Combining operations · Quantum gates

As shown in Chap. 5, utilizing DDs for Schrödinger-style quantum-circuit simulation allows for significant improvements in many cases. While heavily optimized state-of-the-art solutions developed over several years (e.g., those proposed in [64, 86, 145, 157, 168]) required days and/or supercomputers for certain simulations, the initial version of the presented DD-based simulator (cf. Chap. 5) can handle these instances within minutes on a regular Desktop machine. However, further optimizations are possible for DD-based simulation. This chapter (based on [200]) dives deeper into DD-based simulation and discusses in detail that there is still further potential that can be exploited.

As discussed in the previous chapter, DD-based simulation is conducted by multiplying a series of unitary matrices M_i (representing the quantum operations) to a vector v_0 (representing the initial state of the quantum system)—yielding a series of matrix–vector multiplications. However, there is the possibility to combine several operations (requiring matrix–matrix multiplications) before applying it to a vector. But since, from a theoretical perspective, matrix–vector multiplication is significantly cheaper than matrix–matrix multiplication, the potential of this direction was rather limited thus far (approaches such as proposed in [64, 168] used similar techniques but, after all, still suffer from the general limitations of array-based simulators). This chapter (based on [200]) shows contrary findings when employing decision diagrams in simulation. Despite the theoretical complexity, their more

© Springer Nature Switzerland AG 2020

A. Zulehner, R. Wille, *Introducing Design Automation for Quantum Computing*,

https://doi.org/10.1007/978-3-030-41753-6_6

compact representation frequently makes matrix–matrix multiplication more beneficial. Motivated by this, several strategies are discussed that combine operations (using matrix–matrix multiplication) before a simulation step (using matrix–vector multiplication) is conducted. The conducted evaluation confirms that this allows accelerating DD-based simulation of quantum computations by several factors or, by additionally exploiting application-specific knowledge when combining operations, even several orders of magnitude.

In the following, optimizations for DD-based simulation are presented. To this end, Sect. 6.1 discusses the available potential in DD-based simulation resulting from the associativity of matrix multiplication and the fact that the DD-based representation of the considered gate matrices is linear in size. Based on that, Sect. 6.2 presents strategies that exploit this potential. Eventually, Sect. 6.3 evaluates these strategies and compares them to the initial DD-based simulation approach discussed in the previous chapter.

6.1   Potential in DD-Based Simulation

In general, the task of simulating a quantum circuit boils down to multiplying a series of unitary matrices M_i with $0 \leq i < |G|$ (representing a total of $|G|$ quantum operations) to a vector v_0 (representing the initial state of the quantum system). That is, the resulting state vector v_g is typically determined by a series of matrix–vector multiplications:

$$v_{|G|-1} = \left(M_{|G|-1} \times \left(M_{|G|-2} \times \cdots \times (M_0 \times v_0) \cdots \right) \right). \tag{6.1}$$

This simulation could also be done in a different fashion by rearranging parenthesis in Eq. 6.1 (since matrix–matrix multiplication is associative), but requires multiplying matrices with each other. Then, the resulting state vector $v_{|G|-1}$ is determined as series of matrix–matrix multiplications prior to a matrix–vector multiplication, i.e.,

$$v_{|G|-1} = \left(M_{|G|-1} \times M_{|G|-2} \times \cdots \times M_0 \right) \times v_0. \tag{6.2}$$

However, conventionally conducting a matrix–matrix multiplication turns out to be computationally more complex than conducting a matrix–vector multiplication. In fact, even the best known algorithm specialized for multiplying square matrices (cf. [90]) has a complexity of $O(m^{2.373})$, while matrix–vector multiplication has a complexity $O(m^2)$.[1] Because of this, the scheme sketched in Eq. 6.1 is typically utilized.

---

[1]Note that, when considering quantum-circuit simulation as done here $m = 2^n$ (where n is the number of qubits).

But this should change when DD-based simulations are conducted as they offer further potential. Although matrix–matrix multiplication in DDs also requires computing more sub-products and sub-sums per node compared to matrix–vector multiplication (cf. Figs. 5.4 and 5.5), it also has to be taken into account that DDs representing elementary operations usually require significantly fewer nodes (their size is linear with respect to the number of qubits as discussed in Sect. 5.2.1). In contrast, vectors are usually rather complex after the first operations have been applied—leading to rather large DDs. This frequently leads to situations where, using DDs, a matrix–matrix multiplication is cheaper than a matrix–vector multiplication, although more steps per node are required (since re-occurring sub-products only have to be computed once). Thus, multiplying some small (with respect to the number of nodes) matrices first and multiplying the resulting product to a large vector only once (i.e., partially following Eq. 6.2) may be cheaper than sequentially multiplying a large vector to a sequence of small matrices (i.e., following Eq. 6.1).

Example 6.1 *Figure 6.1 shows the DDs representing an intermediate state vector v_i as well as two elementary operations M_i and M_{i+1} (taken from a run that simulates a random circuit as proposed in [16]) as well as the intermediate results when conducting a corresponding simulation following Eq. 6.1 (depicted in Fig. 6.1a) and following Eq. 6.2 (depicted in Fig. 6.1b).[2] As can be clearly seen, following the established simulation flow with the supposedly cheaper matrix–vector multiplications, rather large DDs (i.e., the intermediate state vectors v_i and v_{i+1}) are processed in both of the two multiplications. In contrast, conducting the supposedly more expensive matrix–matrix multiplication first (to combine the two rather small DDs) requires operating on a large DD (representing the vector) only once—significantly reducing the overall computational costs.*

6.2 Exploiting the Potential for More Efficient DD-Based Simulation

The fact that matrix–matrix multiplication might be cheaper than matrix–vector multiplication has not been exploited in DD-based simulation. At the same time, naively relying on matrix–matrix multiplication only (i.e., *completely* following Eq. 6.2) does not necessarily yield an improvement (since, after all, the resulting representation of the intermediate matrices will grow as well). Accordingly, a compromise between the extreme cases (completely following Eqs. 6.1 or 6.2) is required. This leaves the question how many and what operations shall be combined (by matrix–matrix multiplication) before another simulation step (i.e., matrix–vector multiplication) is applied.

---

[2]Note that, for the purpose of this example, it is not essential to provide all the details of the DDs, but it is sufficient to get an intuition of the size of them.

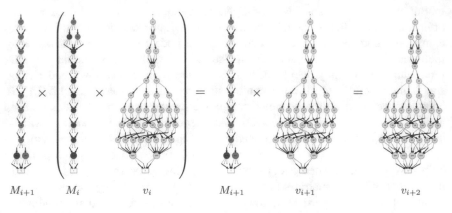

(a) Conducting $v_{i+2} = M_{i+1} \times (M_i \times v_i)$ (i.e., following Eq. 6.1)

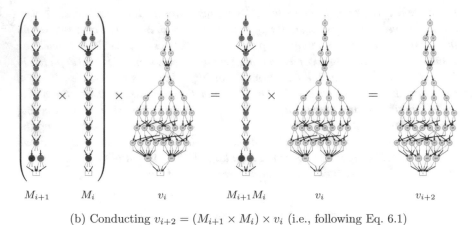

(b) Conducting $v_{i+2} = (M_{i+1} \times M_i) \times v_i$ (i.e., following Eq. 6.1)

Fig. 6.1 Effect of rearranging parenthesis when computing $v_{i+2} = M_{i+1} \times M_i \times v_i$

This section presents solutions to this question. To this end, general strategies are discussed that utilize the potential motivated in Sect. 6.1 as well as strategies, which additionally take further knowledge about the quantum computation to be simulated into account. The conducted evaluation summarized afterwards in Sect. 6.3 demonstrates the substantial impact of these strategies.

6.2.1 General Strategies for Combining Operations

First, general strategies are presented that utilize the potential observed above. To decide how many operations shall be combined, the respective efficiency of DD-based simulation is taken into account. This efficiency usually depends on (1) the fact that the size of a DD representing a product of operations usually grows with the number of its factors (i.e., the number of combined operations) and (2) the fact that the costs of a multiplication heavily depends on the size of the DD representing the matrices and vectors. These observations motivate the following two strategies for combining operations:

- The first one (denoted *k-operations* in the following) forms the product of k operations before multiplying the resulting unitary matrix to the state vector. This directly considers the first observation. However, the size of the resulting product might be very small for one sequences of k operations, but huge for another sequence of k operations.
- The second strategy (denoted *max-size* in the following) avoids this problem and combines operations with respect to the size of the resulting DD. More precisely, operations are combined until their product exceed a certain size (defined by a parameter s_{max}). Then, the resulting matrix is multiplied to the state vector. Accordingly, parametrization is not with respect to just the pure number of operations but with respect to size of the corresponding DD.

Overall, both strategies aim for reducing the number of matrix–vector multiplications at the expense of increasing the number of matrix–matrix multiplications. For simulations where the intermediate state vectors are composed of a large number of DD-nodes, a reduction of the simulation time is expected since these large DDs are involved in fewer multiplications.

6.2.2 Strategies Utilizing Further Knowledge

The two strategies presented above ignore the actual quantum circuit to be simulated and completely rely on generic parameters. However, further potential can be exploited if knowledge about the quantum circuit to be simulated is taken into account.

For example, there exist several quantum algorithms where identical sub-circuits are repeated several times. Grover's algorithm for database search, where a so-called Grover iteration is conducted several times, is a well-known example for this. Figure 6.2 sketches the respective procedure in terms of quantum circuit notation. Here, a database U is concurrently queried with 2^n inputs (achieved by setting n qubits in superposition using an H-operation) and, afterwards, a diffusion operator is repeatedly applied to increase the probability of the desired database entry (the Grover iteration). To maximize the probability of getting the desired database entry, one has to repeat the Grover iteration $2^{n/2}$ times (assuming that there exists exactly one element in the database that satisfies the search criterion)—leading to a quadratic speedup compared to a conventional algorithm.

When simulating such a quantum algorithm, the repeating sequence of operations obviously does not have to be considered completely from scratch each and every time. Instead, this constitutes a perfect sequence to be utilized for combining

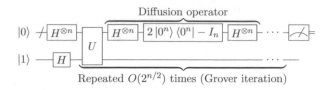

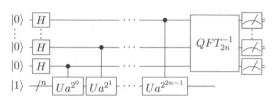

Fig. 6.2 Quantum circuit sketching Grover's algorithm

Fig. 6.3 Quantum circuit
realizing Shor's algorithm

operations. More precisely, rather than considering elementary operations in each iteration, all operations of a single iteration are combined first (yielding a DD representing the entirety of a single iteration). Then, for each further iteration the current state vector has to be multiplied with this combined matrix representation. This does not only save computation efforts because of the effects discussed above (by means of Example 6.1 and Fig. 6.1), but also because the sequence is—once it is pre-computed for the first iteration—reused for all further iterations, without the need of conducting any further matrix–matrix multiplications to combine operations. In the following, this strategy is denoted *DD-repeating*.

Moreover, even if a certain sequence of operations only occurs once, knowledge about the nature of the considered quantum computation may still help. In fact, many quantum algorithms include large Boolean parts (also denoted *oracles*) for which several different realizations (i.e., sequences of elementary operations) exist. Choosing and combining those operations in a fashion which suits DD-based simulation (and not in a fashion given by the quantum circuit) leads to further speedups.

Shor's algorithm [143] is a good example for that. This algorithm translates the problem of factoring an n-bit number N to the problem of determining the multiplicative order r of another number a that is co-prime to N, i.e., r shall satisfy $a^r \equiv 1 \mod N$. To solve this problem on a quantum computer, one can use the circuit shown in Fig. 6.3, where modular exponentiation is performed conditionally on an n-qubit register (the Boolean components Ua^{2^i} compute $x \times 2^{a^i} \mod N$ for an input x). Eventually, an inverse *Quantum Fourier Transform* (QFT) is conducted in order to determine r with high probability.[3]

---

[3]Note that the inverse QFT can also be conducted on a single qubit by using intermediate measurements [13, 63].

When executing this algorithm on a real quantum computer, the Boolean components are *compiled* into elementary quantum operations (cf. Chap. 1) available on the target hardware. This yields rather complex sequences of elementary operations and even requires the addition of further so-called working (or ancillary) qubits. For example, the realization proposed in [63] requires $n + 1$ such working qubits—resulting in a total number of $2n + 2$ qubits for factoring an n-bit number. Instead, without breaking down the oracles, no working qubits and, hence, only $n + 1$ qubits are required. Since it makes no difference for the quality of simulation whether the original functionality of Boolean components or the decomposed version is considered, combining their operations and additionally realizing them in a fashion, which suits DD-based simulation is a promising strategy. In fact, constructing the DD for this functionality not through a sequence of elementary operations but in a direct fashion allows reducing the number of matrix–matrix multiplications significantly and also leads to a smaller number of qubits to be considered (yielding exponential reductions in size). In the following, this strategy is denoted *DD-construct*.[4]

6.3 Evaluation

The strategies presented above exploit further potential in DD-based quantum-circuit simulation and have been implemented on top of the simulator described in Chap. 5, which, hence, serves as baseline.[5] As benchmarks serve several instances of Shor's algorithm [143] (using the implementation provided by Beauregard [13]) and Grover's algorithm [59], since these require larger DDs to represent the state vectors (cf. Sect. 5.4).[6] This section summarizes the conducted evaluation.

First, the results obtained by applying the general strategies as presented in Sect. 6.2.1 are considered. Figures 6.4 and 6.5 provide the speedups obtained when following strategy *k-operations* and strategy *max-size* compared to the original DD-based simulation, respectively. Here, the x-axis indicates the respective values chosen for the parameters k and s_{max} (cf. Sect. 6.2.1), while the y-axis indicates the respectively obtained speedup. Colors indicate the respective benchmark and the interpolated line shows the average speedup obtained for each value of k and s_{max}.

These results confirm the discussions conducted in Sect. 6.1: Conducting simulation using only matrix–vector multiplications (i.e., following Eq. 6.1) does not fully utilize the possible potential. Instead, combining operations (using the supposedly

---

[4]Note that a similar scheme called *emulation* has been discussed in [64] for array-based simulation, where word-level arithmetic functions allow simulating modular exponentiation directly.

[5]Chapter 5 provides a general comparison of DD-based simulation to other simulation approaches.

[6]In the following, those benchmarks are denoted by *shor* and *grover*, respectively, followed by a number indicating the number of qubits in the computation. For Shor's algorithm, the name also includes the number N to be factored as well as the number a co-prime to N, since these numbers significantly affect the simulation time. That is, those benchmarks have the form *shor_N_a_qubits*.

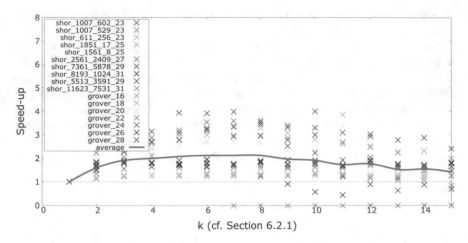

Fig. 6.4 Speedup for strategy *k-operations*

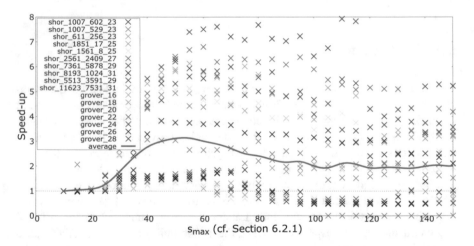

Fig. 6.5 Speedup for strategy *max-size*

more expensive matrix–matrix multiplications) to a certain extent (i.e., setting k or s_{max} to a value, which allows for a combination of operations) improves the runtime significantly. The results, however, also confirm that combining *all* operations (i.e., completely following Eq. 6.2) still is not a suitable option. At some point, the benefits sketched in Example 6.1 and Fig. 6.1 disappear, and the DD representing the resulting matrix gets too large. Overall, speedups of up to a factor of 3.99 (for *k-operations*) and 7.93 (for *max-size*) are observed in the best cases.

The second part of the evaluation shows the improvements gained by utilizing further knowledge as presented in Sect. 6.2.2. As discussed there, this is applicable for the *grover*-benchmarks (in case of strategy *DD-repeating*) and the *shor*-benchmarks (in case of *DD-construct*). Tables 6.1 and 6.2 list the correspondingly obtained results. The first column provides the name of the benchmarks, the second column the run-time of the DD-based approach described in Chap. 5, and the third one the run-time obtained by the best choice of k or s_{max}. In addition to that, the forth column additionally gives the required run-time when the respective strategy is applied.

As can be seen, the strategies discussed in Sect. 6.2.2 yield substantial improvements. In case of the *grover*-benchmarks (additionally reusing a DD representing a combined sequence of operations after the first iteration), further speedups up to a factor of almost 2 can be achieved. In case of the *shor*-benchmarks (additionally utilizing a better suited DD construction, and particularly fewer qubits to represent), further speedups of several orders of magnitude are achieved. More precisely, this strategy frequently allows boiling down the run-time from minutes or hours to just very few seconds or even less.

Overall, for a DD-based simulation approach, which already has shown its advantages compared to other established simulators such as [64, 86, 145, 157, 168], substantial further potential could be identified and utilized.

Table 6.1 Results for *grover*-benchmarks (strategy *DD-repeating*)

Benchmark	t_{sota}	$t_{general}$	$t_{DD\text{-}repeating}$
grover_16	0.92	0.49	0.08
grover_18	2.55	1.26	0.84
grover_20	6.15	3.45	4.05
grover_22	19.28	8.96	9.16
grover_24	54.31	21.99	19.61
grover_26	131.43	56.57	40.26
grover_28	330.90	141.86	87.83

Table 6.2 Results for *shor*-benchmarks (strategy *DD-construct*)

Benchmark	t_{sota}	$t_{general}$	$t_{DD\text{-}construct}$
shor_611_256_23	77.31	13.45	0.13
shor_1007_529_23	23.01	7.87	0.04
shor_1007_602_23	121.05	19.41	0.21
shor_1561_8_25	93.34	18.36	0.12
shor_1851_17_25	172.23	28.42	0.29
shor_2561_2409_27	374.04	55.95	0.48
shor_5513_3591_29	1050.25	131.54	1.00
shor_7361_5878_29	258.50	44.41	0.28
shor_8193_1024_31	104.79	28.17	0.13
shor_11623_7531_31	2426.34	467.39	4.01

Chapter 7
Efficient Implementation of the DDs in the Quantum Realm

Abstract This chapter provides implementation details of the underlying DD-package. While many concepts are directly taken from conventional DD-package, the occurring complex numbers cause significant issues—especially since they occur as weights attached to the edges of the DD and since limited machine accuracy inevitably leads to numerical errors. This chapter discusses concepts to handle complex numbers more efficiently and, hence, yields an efficient DD-package for the quantum realm that achieves improvements of several orders of magnitude compared to previous implementations. This does not only affect the proposed simulation approach, but can potentially improve other DD-based design methods (e.g., for synthesis or verification) as well, just by exchanging the underlying DD-package.

Keywords Quantum computing · Decision diagrams · DD-package · Efficient implementation · Complex numbers · Lookup table

The previous two chapters discussed the conceptual foundation of DD-based simulation. The conducted evaluation showed significant improvements compared to straightforward solutions for many practically relevant benchmarks, since the state vector is represented much more compactly in the presented DD-based approach—outweighing the constant overhead caused by the graph structure.[1] However, it remains obvious that the enormous improvements as listed in Table 5.1 also depend on an efficient implementation of the underlying DD-package—similar to the conventional domain, where the design automation community successfully addressed many challenges by the introduction of decision diagrams and corresponding efficient implementations (cf. Sect. 3.1) that affect the development of design tools and methods until today.

---

[1]Note that DDs for the quantum realm have also been utilized for tackling the exponential complexity in tasks like synthesis [116, 117, 119] and verification [118, 162, 167, 181]—relying on decision diagram types as proposed in [3, 104, 115, 164, 167].

© Springer Nature Switzerland AG 2020
A. Zulehner, R. Wille, *Introducing Design Automation for Quantum Computing*,
https://doi.org/10.1007/978-3-030-41753-6_7

Key concepts required for implementing efficient DD-packages—such as unique tables, garbage collection with reference counts, or compute tables—are known from the conventional domain (which, also in the 1990s, have been explicitly investigated, e.g., in [21, 23, 67, 102, 158]) and, hence, are directly incorporated. However, the quantum realm additionally requires an efficient handling of complex numbers. They introduce several new problems such as how to keep numerical stability, how to efficiently store nodes in unique tables, how to store reoccurring operations in compute tables, as well as several further issues that have not been considered thus far.

This chapter (based on [186]) presents details on how to efficiently implement a DD-package for the quantum realm addressing these arising problems. Concepts from decision diagrams for the conventional domain are reused where applicable, while approaches for tackling problems caused by focusing on quantum computations are described in detail. An evaluation confirms that these efficient implementation techniques yield improvements of orders of magnitude with respect to run-time compared to the best known implementation available today. An implementation of the resulting DD-package is publicly available at http://iic.jku. at/eda/research/quantum_dd.

In the following, the implementation details are provided. To this end, Sect. 7.1 reviews concepts known from DD-packages from the conventional domain and discusses the difficulty of handling complex values. Based on that discussion, Sect. 7.2 describes methods to handle the occurring complex numbers efficiently. Eventually, Sect. 7.3 presents the resulting DD-package for the quantum realm and provides a comparison to previously developed packages.

7.1 General Concepts and Challenges

This section discusses how to implement DD-packages for the quantum domain in an efficient fashion. To this end, concepts such as *unique tables*, dedicated *garbage collection*, or *compute tables*—techniques which are already taken for granted in DD-packages for the conventional domain [66, 88, 156]—are utilized. However, this section also shows that just adopting these concepts is not sufficient for a highly efficient quantum DD-package and that additional implementation techniques are required which allow for an efficient handling of complex numbers and sub-factors thereof (essential for storing the edge weights discussed in Chap. 5).

7.1.1 Established Implementation Techniques

In the 1990s, key concepts for efficient implementations of DD-packages for the conventional domain have been introduced [21, 23, 67, 102, 158], which leveraged the development efficient realization of tasks in verification or synthesis (using DD-packages such as [66, 88, 156]). Many of these concepts apply directly to decision diagrams in the quantum domain, namely:

- *Unique tables* which store the nodes of the decision diagram and allow determining redundancies in the structure efficiently. The *unique table* is realized as 2-dimensional hash table (one for each variable in the decision diagram), where each bucket of the table contains a linked list of DD-nodes that are associated with the same hash value and variable. Before any new DD-node is inserted, it is checked whether the node already exists in the *unique table*.[2] A DD-node is returned by means of a pointer to the respectively allocated memory to allow dereferencing in constant time. Hence, it can be denoted as *strong canonical form* [21], i.e., a unique representation.
- Dedicated *garbage collection* which frequently removes unused DD-nodes from the unique table and, by this, allows for a fast insertion of new nodes as well as for keeping the memory usage low. To this end, *reference counting* is used to keep track which node shall be kept in the unique table. In order to avoid recurring memory allocations and deallocations, free DD-nodes (i.e., allocated nodes that are currently not stored in the unique table) are stored in a list and corresponding memory is allocated in blocks with capacity of storing several list entries.[3] Nodes that are removed from the unique table are appended to this list. To reduce the overhead caused by the lists (the list of free nodes as well as the lists in the buckets of the unique table), DD-nodes themselves contain a pointer to the next element in the list.
- *Compute tables*, which cache the results of operations that are repeatedly conducted on the same DD-nodes. Since decision diagrams exploit redundancies by using shared nodes, this proves very beneficial as many operations are frequently repeated and, hence, do not have to be recomputed. Compute tables are realized as hash tables, where the DD-nodes denoting the operands (and possibly the type of operation) are used to determine the hash key [21]. The entries in the hash table might contain a single result that may be overwritten, or linked lists to store all computations with the same hash key to increase the *hit rate* (e.g., with an LRU strategy as proposed in [158]). Note that the compute tables are (partially) invalidated/cleared after garbage collection, since operands (referenced by a pointer) may be overwritten after the respective nodes have been removed from the unique table.

In fact, all these concepts can be (and have been) directly incorporated into implementations of decision diagrams for quantum computation described in Sect. 5.1. However, decision diagrams for quantum computations additionally have to handle the frequently occurring complex numbers. This constitutes *the* major

---

[2]The hash function should distribute the nodes equally to the buckets to reduce the number of collisions at lookup.

[3]Whenever the list is empty, enough memory is allocated to instantiate several new DD-nodes.

obstacle towards an efficient implementation of a fully fletched DD-package for the quantum domain as discussed in the next section.

7.1.2 Handling Complex Numbers

As discussed in Sect. 5.1, weights that are attached to DD-nodes offer the possibility for further compaction. In the conventional domain, having such edge weights does not constitute an issue from the implementation perspective, since they are (tuples of) integers [24, 66]—again, a strong canonical form (a unique representation). This allows using the efficient concepts outlined above, since computing unambiguous hash keys for DD-nodes (containing weights attached to outgoing edges) is still possible.

In the quantum domain, however, weights are formed out of complex numbers. From a mathematical perspective, this causes no problems since complex numbers also provide a strong canonical form. However, it introduces severe challenges from an implementation perspective where machine accuracy is limited and, hence, complex numbers are approximated—yielding to numerical errors in computations. In fact, changing one complex number attached as weight to an outgoing edge by a tiny fraction (e.g., by flipping the least significant bit of the mantissa of the real or imaginary part) may yield a completely different hash key. Accordingly, redundancies might remain undetected by the DD-package since the node is searched in the wrong bucket of the unique table—causing a substantially larger decision diagrams even though redundancies are actually present. Additionally, since weights represent sub-factors of complex numbers, this further increases the probabilities of corresponding numerical instabilities.

To overcome this issue, one can represent complex numbers as two quadratic irrational numbers of the form $\frac{a+b\sqrt{2}}{c}$ with $a, b, c \in \mathbb{Z}$ for the real and imaginary part [105, 163]—limiting the set of representable complex numbers and, thus, not allowing for representing arbitrary quantum computations. Alternatively, one can use an additional table to store complex numbers, where an edge weight is then represented by an index of this table holding the corresponding number[121, 163]. The table is then maintained in a fashion such that numbers differing by not more than a tolerance value ϵ share an entry in this table.[4]

However, while using a lookup table for complex numbers indeed allows for representing arbitrary quantum computations, it introduces several new problems when aiming for an efficient implementation. Thus, rather straightforward implementations of the respective DD-concepts are available only. The remainder of this chapter identifies arising difficulties and presents implementation techniques, which, explicitly allow for implementing DDs with complex-valued edge weights in an efficient fashion.

---

[4]Note that, in [163], this ϵ is a relative tolerance value, while it is absolute in [121].

7.2   Efficient Handling of Complex Edge Weights

As discussed above, handling complex-valued edge weights (and complex numbers in general) in an efficient fashion is a key to a fully fletched DD-package for quantum computing. This section presents such techniques—covering arising issues like numerical instabilities caused by ϵ, how to realize an efficient lookup of complex numbers that considers ϵ (thus providing a strong canonical form), as well as how to handle operations on DDs efficiently.

7.2.1   Obtaining Numerical Stability

Simulation as well as other design tasks for quantum computing (like synthesis or verification) heavily rely on multiplying unitary matrices either with each other or with vectors. From a numerical perspective, these operations are not critical, since the multiplication with a unitary matrix is a well-conditioned operation—even when the individual entries of the matrix are determined as product of several factors (as done in an edge-valued decision diagram). However, this changes when introducing the tolerance value ϵ as discussed above to find redundancies. More precisely, some factors of an entry in the unitary matrix might be significantly rounded.

This problem becomes evident when using a normalization scheme (to gain canonicity of DD-nodes) as described in Sect. 5.1. Here, all weights of outgoing edges are simply divided by the leftmost non-zero outgoing edge weight while propagating this extracted factor to the parent nodes—edge weights in the decision diagrams are likely to become either rather large or rather small. If the real and the imaginary part of an edge weight are now close to 0 (i.e., in the interval $[-\epsilon, \epsilon]$), the weight is rounded to 0. By this, a subset of the DD-nodes possibly vanishes by setting several entries in the matrix (vector) to zero—ending up with a huge round-off error and numerical instabilities.

A better numerical stability is reached by changing the utilized normalization scheme—by dividing all weights of the outgoing edges of a DD-node by the weight with the largest absolute value (squared magnitude). If several outgoing edges have attached weights with equal absolute values (as can be seen in Fig. 5.2a on Page 41 for the node labeled q_0), the leftmost of these weights is chosen to preserve canonicity. With this normalization scheme, it is guaranteed that all edge weights in the decision diagram have an absolute value between 0 and 1 (and, thus, also the absolute values of the real and imaginary parts are between 0 and 1)—making it less likely that a factor unintendedly rounds to 0. While this indeed helps to increase numerical stability, one can additionally exploit the knowledge that all occurring complex numbers are either on or inside the unit circle to store them efficiently. This is discussed in the next section.

7.2.2 Looking-Up Complex Numbers

As discussed in Sect. 7.1.2, hashing complex numbers is not possible due to rounding errors caused by the limited machine accuracy. Hence, different methods are required that allow for a unique and efficient lookup of complex numbers while still considering the tolerance value ϵ.

The general idea for an efficient lookup is to exploit that numbers can be sorted. Since there does not exist a total order for complex numbers, they are split into their real and imaginary parts. These real-valued parts are then stored separately in a lookup table.[5] A complex number is then represented by a pair of pointers to elements in the lookup table—a strong canonical form. Moreover, storing only the absolute value of the real and imaginary part and "hide" the sign bit in the pointer to the respective entry of the table allows conducting operations like multiplication with constants like -1, i, or $-i$, as well as computing the complex conjugate just by flipping bits and/or swapping pointers.

The realization of a lookup table for real-valued entries exploits the knowledge of the normalization scheme discussed above. In fact, it is guaranteed that all numbers of the table are within the interval $[0, 1]$.[6] An efficient lookup is achieved by splitting this interval into N equally distributed chunks.[7] These chunks are represented as entries in an array of size N, where each entry initially contains an empty list for occurring numbers in the respective interval (similar to the buckets in a hash table). When a new real number r shall be inserted, the corresponding bucket is traversed. If one of the numbers in this list is equal to r (considering the tolerance value ϵ), a pointer to the respective number is returned. Otherwise, r is inserted into the bucket.

However, one has to be careful since, by allowing a tolerance value of ϵ, a sufficiently close number might be located in one of the neighboring buckets. More precisely, if $r - \epsilon$ or $r + \epsilon$ exceeds the border of the considered interval (represented by a bucket), one has to additionally look for a number sufficiently close to r in the corresponding neighboring bucket. For performance reasons, the first value which deviates less than ϵ from r is returned. Alternatively, to improve numerical stability, one can continue searching for a value that is even closer to r.[8]

The implementation techniques used for the unique table are also employed for the lookup table to allow for an efficient realization. This includes reference counting to keep track of which entries of the lookup table are still required, a list of "free" numbers, as well as a procedure to allocate memory for several new entries at once. Whenever garbage collection is conducted on the unique table, another garbage collection routine runs on the lookup table to remove entries (appending

---

[5]Note that a separate insertion of the real and imaginary part may also reduce the overall numbers to be stored since the same number might occur as real/imaginary part in several complex numbers.

[6]Note that this also holds for the real and imaginary part of the complex weight attached to the edge pointing to the root node since the DD represents a unitary matrix.

[7]Note that other strategies for splitting the interval $[0, 1]$ are possible.

[8]Note that there exist at most two numbers in the lookup table that are closer than ϵ to r.

them to the list of free entries) with a reference count of 0. Moreover, numbers can be easily dereferenced since pointers are returned.

By using a lookup table as discussed above, inserting a value r has complexity $\mathcal{O}(1)$ if N is chosen suitably and the (inserted) numbers are distributed equally in the interval $[0, 1]$ (to avoid long collision chains). However, like a hash table, a worst-case complexity of $\mathcal{O}(n)$ results when all n entries in the table are stored in the same bucket. As alternative to a table, one can also use a self-balancing binary tree like an AVL tree [134] or a red-black tree [60] to store the real numbers. Then, the lookup of a number r requires $\mathcal{O}(\log n)$, where n is the number elements in the tree. This is again possible, since real numbers can be totally ordered, but additionally requires overhead for re-balancing the tree. Eventually, a combination of both—a lookup table where all entries stored in one bucket are realized as a self-balancing binary tree—providing a compromise of both ideas from a complexity-theoretic consideration.

However, even with an efficient implementation of a lookup for complex (real) numbers, several other issues have to be considered when conducting operations on decision diagrams. These are discussed in the following section.

7.2.3   Conducting Operations on DDs

This section discusses how to efficiently handle complex-valued edge weights when conducting operations on decision diagrams. To this end, arising issues (from using a lookup table for complex numbers) that might impair efficiency are analyzed first, namely:

- Each intermediate computation on complex numbers requires performing a lookup in the table and, additionally, may cause rounding. Such intermediate values occur before normalizing a DD-node.
- Sub-results might contain complex numbers with absolute value larger than 1 (before normalizing a DD-node).

The first issue may affect the efficiency since the number of entries in the lookup table grows significantly (by inserting intermediate values that are not used anymore afterwards). Moreover, intermediate computations are not conducted as accurately as possible (by performing a lookup that considers ϵ). The second issue even prohibits the use of a lookup table as introduced above, since it is not guaranteed that all entries are within interval $[0, 1]$. However, both issues are resolved by introducing a cache for complex numbers that are used as intermediate results. Entries are taken from this cache whenever intermediate results are computed and fed back when normalizing a newly computed DD-node (before looking it up in the unique table).

In general, the efficiency of operations on decision diagrams results from a recursive formulation. Repeatedly re-evaluating the same (sub-)operations is avoided by using a compute table. Since each recursive call returns an already normalized DD-node, cached complex numbers are only required for the current recursion level. Moreover, also the recursion levels above the current one may hold some cached complex numbers representing sub-results from other recursive calls (at most one for each other outgoing edge). This implies that a cache size linear to the number of variables is sufficient. Even more, by fixing the maximum number of variables in the decision diagram beforehand, a cache with fixed size for complex numbers (which is allocated at initialization of the package) is sufficient.

This cache is implemented as list of real numbers, which utilizes the same data structure as for entries in the lookup table discussed above. This allows using entries from the lookup table and the cache interchangeably when computing, e.g., the product of two complex numbers. Cached complex numbers are allocated by taking two real numbers from the front of the list representing the cache. Before feeding them back to the cache (when normalizing the computed DD-node), they are inserted into the lookup table (if no suitable number has been inserted yet). This is necessary since the lookup in the unique table requires a strong canonical form for the weights attached to the outgoing edges to detect redundancies. Since complex numbers are looked up as late as possible, all computations are conducted with maximal precision. Intermediate results are not inserted into the lookup table for complex numbers, which keeps the number of entries in this table low and avoids repeated rounding (caused by ϵ) during computations.

On the downside, having a cache for complex numbers affects the compute tables that store sub-results. In fact, depending on the considered operation, an operand's weight is either stored in the cache or in the lookup table. Complex numbers stored in the cache again have the problem that they do not necessarily preserve a strong canonical form. Hence, a completely different hash key may result from two slightly different complex numbers. This is not as critical as in the unique table, since only the hit rate of the compute table may decrease. However, an internal evaluation has shown that the hit rate is hardly affected when rounding the complex numbers before computing the hash key.

7.3 Resulting DD-Package

This section describes the resulting DD-package for quantum computing when utilizing the implementation techniques introduced above. As baseline serves the QMDD-package provided at http://informatik.uni-bremen.de/agra/eng/qmdd.php, which already utilizes the implementation techniques common for DD-packages in the conventional domain such as unique and compute tables, as well as garbage collection using reference counting. However, complex numbers are handled in a straightforward and inefficient fashion, namely:

- Complex numbers are stored with an array of fixed size.
- The array is linearly traversed at each lookup, i.e., inserting a new number forces the traversal of the complete array.

- Complex numbers are inserted into the array at each step (causing rounding of intermediate results, which may affect numerical stability).
- The normalization scheme does not consider numerical implications caused by ϵ.

All these shortcomings make the original QMDD-package a proof of concept implementation (handling only small instances efficiently) rather than a fully fletched DD-package for quantum computation.

The implementation techniques for an efficient handling of complex values (presented in this chapter) address these shortcomings and, hence, have been implemented on top of this package in C. The resulting DD-package is publicly available at http://iic.jku.at/eda/research/quantum_dd.

To demonstrate the improved efficiency of the developed package, the corresponding decision diagrams of several established quantum computations have been generated. More precisely, the *Quantum Fourier Transform* (QFT [45]) and random circuits proposed by Google to conduct quantum supremacy experiments [16] serve as representatives. Since building the decision diagrams representing the respective functionalities is conducted by successively multiplying the individual quantum gates, this serves as a representative case study on the efficiency of the package (after all, design automation methods heavily rely on matrix multiplication).[9] The timeout was set to one hour.

Table 7.1 provides the obtained results by listing the number of qubits q, the number of quantum operations #*op*, the *size* (i.e., the number of DD-nodes) of the resulting decision diagram, as well as the number of different complex values #*complex* occurring throughout the computation.[10] Moreover, the table lists the runtime for building the decision diagram representing the respective functionality when using the originally available implementation (denoted $t_{original}$) as well as when using the improved DD-package utilizing the implementation techniques presented in this chapter (denoted $t_{improved}$).[11]

The obtained results clearly show that the presented techniques allow handling complex numbers much more efficiently than the original implementation. For example, consider the random circuits (proposed in [16]) with 16 qubits. Building the DD for the quantum functionality given by the first 50 operations (gates) is handled by both packages in a fraction of a second. Considering the first 70 operations already results in a DD with more than 6000 nodes and requires dealing with close to 20,000 complex numbers. Even though these are not large numbers for decision diagrams, the original package already requires several seconds to build the decision diagram. When considering the first 75 operations, dealing with

---

[9]Simulation was not considered as dedicated application since the shortcomings of the original package often prevent conducting simulation in reasonable time—even for small instances.

[10]Note that the number of different complex number can only be determined for the original package since all of them remain in an array during the computation.

[11]Note that the original package was slightly adjust to allow for storing more than 10,000 different complex numbers. This, however, did not affect the run-time performance of the package.

Table 7.1 Efficiency of the presented implementation techniques

Name	q	#op	Size	#complex	$t_{Original}$	$t_{Improved}$
Supremacy	16	50	80	238	0.00	0.00
Supremacy	16	70	6,278	18,803	4.04	0.06
Supremacy	16	75	36,161	225,560	1147.14	0.57
Supremacy	16	80	1,195,979	–	TO	65.18
Supremacy	20	70	455	3534	0.06	0.01
Supremacy	20	80	899	8010	0.26	0.02
Supremacy	20	89	71,105	97,942	1453.63	2.52
Supremacy	20	95	1,742,795	–	TO	912.11
Supremacy	25	100	912	6151	0.15	0.02
Supremacy	25	110	1751	12,851	0.46	0.03
Supremacy	25	119	58,939	96,647	1282.28	2.63
Supremacy	25	120	365,643	–	TO	3.51
QFT	15	120	32,767	32,785	3.98	0.39
QFT	16	136	65,535	65,554	17.48	0.86
QFT	17	153	131,071	131,091	67.02	1.64
QFT	18	171	262,143	262,164	295.48	3.50
QFT	19	190	524,287	524,309	1269.35	8.38
QFT	20	210	1,048,575	–	TO	22.96

q: number of qubits, *#op*: number of operations, *size*: size of the DD.
#complex: number of occurring complex numbers, $t_{original}$: time when using the original QMDD-package.
$t_{improved}$: time when incorporating the techniques presented in this chapter. The timeout was set to 1 h.

more than 36,000 DD-nodes and with approximately 225,000 complex numbers requires almost 20 min—a task that is solved in less than a second by the DD-package developed in this chapter. In fact, the limiting factor is not only the number of nodes, but also the number of complex values to deal with. Hence, the presented implementation techniques allow building up the functionality of the first 80 operations—a rather large decision diagram with almost 1.2 million nodes—in a bit more than one minute, whereas the original implementation fails to do that within an hour.

Overall, improvements in the performance of several orders of magnitude are observed. Hence, the evaluation demonstrates that complex values (especially when used as edge weights) are handled much more efficiently by using the implementation techniques described in chapter. The resulting DD-package allows for an efficient implementation of the DD-based simulation approach presented in Chaps. 5 and 6, and, hence, substantially contributes to the achieved speedups (compared to array-based simulators).

Chapter 8
Accuracy and Compactness of DDs in the Quantum Realm

Abstract This chapter provides an algebraic representation (related to the Clifford+T library) for the complex numbers occurring in the DDs of the proposed DD-based simulator. This does not only allow simulating Clifford+T circuits in an exact fashion using DDs, but also to evaluate—for the first time—the existing trade-off between accuracy and compactness of decision diagrams in the quantum realm.

Keywords Quantum computing · Decision diagrams · Accuracy · Compactness · Complex numbers · Algebraic representation

The previous chapters have presented a new DD-based simulation approach that—together with an efficient implementation of the underlying DD-package and corresponding optimizations—allows for significant improvements compared to previous approaches in many cases. However, as already discussed in Chap. 7, rounding errors may significantly harm DD-based representations since redundancies might not be detected. More generally, current decision diagrams for the quantum domain (not limited to their application in simulation) suffer from a trade-off between accuracy and compactness:

- On the one hand, small errors are inevitably introduced by the limited precision of floating-point arithmetic on conventional computers. In fact, the complex numbers in the state vectors and transformation matrices often have *irrational* imaginary or real parts that can only be approximated. If these possible errors are not taken into account, redundancies might not be recognized—harming the compactness (i.e., the size of the decision diagram) significantly. That is, a small amount of inaccuracy has to be tolerated in order to find redundancies and, thus, achieve compactness.
- If, on the other hand, too much inaccuracy is tolerated, information is lost which introduces numerical instabilities that—in the worst case—will falsify the results. While for some applications a moderate error may be acceptable (e.g., since the underlying quantum algorithm is robust enough), others require a rather accurate representation.

© Springer Nature Switzerland AG 2020

A. Zulehner, R. Wille, *Introducing Design Automation for Quantum Computing*,
https://doi.org/10.1007/978-3-030-41753-6_8

Consequently, an application-specific trade-off between accuracy and compactness needs to be conducted in order to obtain efficient and sufficiently accurate methods on a case-by-case basis (e.g., for applications like simulation as discussed above, synthesis [116, 117, 119], or verification [118, 162, 167, 181]). Even more, a time-consuming fine-tuning of the corresponding parameters might be necessary in order to adapt design methods to a certain functionality or algorithm. However, as confirmed by thorough analysis of this issue that will be conducted in this chapter, it is not guaranteed that the desired accuracy or compactness is achievable at all, which—in the worst case—might cause corrupted results or infeasible run-times, respectively. This motivates the need for an alternative solution that inherently achieves accuracy *and* compactness at the same time and, thus, allows overcoming the trade-off present in current solutions.

This chapter (based on [120, 187]) presents such an alternative approach that utilizes an algebraic number field (within the complex numbers) that is strongly connected with the well-established description of quantum computations in terms of the *Clifford+T* gate library. This allows representing the considered quantum functionality algebraically rather than numerically—thereby completely avoiding the trade-off between accuracy and compactness. In fact, the utilized decision diagram fully exploits redundancies for a compact representation and, at the same time, guarantees a perfectly accurate result.

In the following, the presented alternative approach is presented. To this end, Sect. 8.1 discusses the trade-off between accuracy and compactness in DDs for the quantum realm. Based on that, Sect. 8.2 presents an algebraic solution to overcome this trade-off. Eventually, Sect. 8.3 evaluates the existing trade-off between accuracy and compactness in DDs for the quantum realm, as well as the efficiency of the presented algebraic approach.

8.1 Trade-Off Between Accuracy and Compactness

While gaining a compact representation without information loss is not complicated as long as the set of possible values is finite (e.g., 0 and 1 for Boolean functions) or discrete (e.g., only integer numbers occur), this is different for the quantum domain, where arbitrary complex numbers and, thus, irrational coefficients occur. Consequently, decision diagrams that represent conventional computations do not face problems with the accuracy of the representation, while accuracy can be an important issue for decision diagrams representing quantum computations (cf. Chap. 7). This section discusses the issues that lead to a trade-off between accuracy and compactness—motivating the need for an algebraic representation of quantum computations overcoming this.

To this end, first recall that, in the area of quantum computing, most design automation tasks require hundreds or even thousands of multiplications of unitary matrices (e.g., to compute the unitary matrix for an entire quantum circuit from the gate matrices) or multiplications of a vector and a unitary matrix (e.g., to simulate the evolution of a quantum state). From a numerical perspective, these tasks do not constitute an issue per se, since the multiplication with a unitary matrix is a well-conditioned problem. In fact, the resulting error, i.e., the deviation from the exact result, is expected to be in the order of the input error.[1] Furthermore, applying several multiplications successively will only lead to an error that grows linearly with the number of matrix multiplications. Consequently, using a numerical, i.e., approximated representation of the complex numbers with a high resolution may yield numerically stable computations.

However, this approximation may significantly affect the compactness of the decision diagram representation. To this end, recall that the key idea of decision diagrams is to exploit redundancies in order to gain a compact representation. This compact representation is indeed a key factor for their efficiency, since the complexity of the manipulation algorithms (e.g., matrix multiplication) grows with the size of the decision diagram. By using approximate representations of the complex numbers, the detection of these redundancies can become a tough challenge. An example demonstrates the problem.

Example 8.1 *Consider the DD representing a Hadamard gate (cf. Fig. 5.2). A compact representation is gained since several redundancies are exploited. Representing the irrational entries of the corresponding matrix with floating point numbers on a machine with limited accuracy may break these redundancies, e.g., when using rounding towards ∞ or when the matrix is constructed as the product of several other matrices. Then, two occurrences of $\pm\frac{1}{\sqrt{2}}$ might be represented by slightly different floating point numbers (differing in a few of the least significant bits of the mantissa) and no redundancies are detected anymore.*

In general, this will likely lead to a matrix or a vector where no redundancies are detected at all—leading to an exponentially large representation. A solution to this issue that some of the redundancies (that actually exist in the matrix) are not detected due to tiny errors caused by the limited machine accuracy is to identify numbers that do not differ by more than a so-called *tolerance value* (denoted as ϵ in the following; cf. Chap. 7).

Example 8.1 (Continued) *Assume that two entries that shall represent $\frac{1}{\sqrt{2}}$ differ only in the last three bits of the mantissa (assuming an IEEE 754 single precision floating point number with 23 mantissa bits). Then, setting, e.g., $\epsilon = 10^{-5}$ allows detecting that the two entries are equal.*

However, choosing a proper value for ϵ is crucial. If ϵ is chosen too small, it might not be able to compensate the limited machine accuracy and, thus, to determine more redundancies. If ϵ is chosen too large, this might lead to numerical instabilities of the multiplication algorithm. Moreover, additional redundancies might be detected that are not actually present—leading to an undesired approximation and, thus, information loss. In the worst case, this may falsify the result

---

[1]Note that this is a statement about the matrix multiplication problem itself and not about a certain implementation.

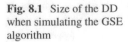

Fig. 8.1 Size of the DD when simulating the GSE algorithm

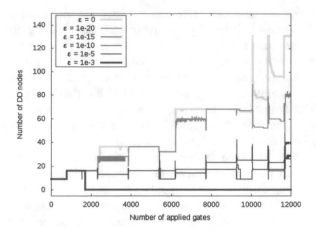

such that an invalid quantum state (e.g., a vector composed of zeros only) or a non-unitary matrix results. Nevertheless, in many cases there exist proper configurations for ϵ, but this heavily depends on the considered application and determining an adequate tolerance value may require time-consuming fine-tuning of parameters on a case-by-case basis.

Example 8.2 *Figure 8.1 shows the size of the DD through simulating the* Ground State Estimation *(GSE, [169]) quantum algorithm. As can be seen, the number of DD-nodes is highly affected by ϵ. Choosing $\epsilon = 0$, i.e., (almost) no two different numbers are considered to be equal yields the highest precision that is possible using floating point numbers, but results in a rather large representation. Instead, choosing $\epsilon = 10^{-3}$ yields a vector composed of zeros only—a perfectly compact but obviously wrong representation. Of course, both choices (highlighted in bold in Fig. 8.1) represent extreme cases. As a trade-off, choosing $\epsilon = 10^{-15}$ leads to almost the same numerical result as $\epsilon = 0$, but yields a better compactness and, thus, a shorter run-time.*

Overall, determining the "perfect" ϵ, i.e., finding the best trade-off between accuracy and compactness (which heavily influences the run-time) is a non-trivial task. So far, it has to be evaluated on a case-by-case level for each application.

This chapter overcomes this trade-off by using an algebraic representation of the complex numbers that occur in the vectors/matrices—eventually resulting in a decision diagram that detects all existing redundancies and computes the result in an exact fashion (i.e., without a numerical error).

8.2   Algebraic Simulation

This section presents a solution for the algebraic representation of complex numbers in DDs, which overcomes the approximation drawbacks caused by the numerical number representation and allows for both, a perfect accuracy together with a perfect exploitation of redundancies. To this end, the properties of the ring $\mathbb{D}[\omega]$ that will be utilized for the exact, algebraic representation of complex numbers are discussed first. After that, a solution for exploiting the benefits of this representation in DDs is presented.

8.2.1 Utilizing the Ring $\mathbb{D}[\omega]$

In order to obtain an algebraic representation of the complex numbers, the most obvious choice would be to extend the well-known Gaussian numbers $\mathbb{Z}[i]$ to the ring $\mathbb{Z}[i, \sqrt{2}]$. By doing so, all complex numbers of the form $a + b\sqrt{2} + i(c + d\sqrt{2})$ can be represented exactly. This ring is already a dense subset of the complex numbers such that any complex number can be approximated by an element from $\mathbb{Z}[i, \sqrt{2}]$ up to an arbitrary precision (this *density* is a known property of $\mathbb{Z}[\sqrt{2}]$ in the real numbers and can be easily lifted to the complex numbers). However, the irrational number $\frac{1}{\sqrt{2}}$ that plays a vital role in quantum computation is not contained in this ring.[2] Thus, it seems more promising to study the ring $\mathbb{Z}[i, \frac{1}{\sqrt{2}}]$ which trivially contains $\mathbb{Z}[i, \sqrt{2}]$ (since $\sqrt{2} = 2 \cdot \frac{1}{\sqrt{2}}$), but allows representing $\frac{1}{\sqrt{2}}$ and all its potencies exactly.

In the following, a different interpretation of this ring is used that is more convenient from an algebraic perspective. More precisely, the interpretation as an extension of the so-called *dyadic fractions* $\mathbb{D} = \{\frac{a}{2^k} \mid a, k \in \mathbb{Z}, k \geq 0\}$, namely $\mathbb{D}[\omega]$ (where $\omega = \frac{1+i}{\sqrt{2}} = e^{i\pi/4}$) is utilized.[3]

Using the latter representation, all complex numbers that can be represented exactly can be written as $\alpha = \frac{1}{\sqrt{2}^k}(a\omega^3 + b\omega^2 + c\omega + d)$ for coefficients $a, b, c, d, k \in \mathbb{Z}$, i.e., using five integers (cf. [53]).

Note that the ring $\mathbb{D}[\omega]$ is also strongly related to the well-established Clifford+T gate library [19]. This library is very popular in quantum computation due to its universality as well as fault-tolerance (cf. Sect. 2.1). The relation between the ring $\mathbb{D}[\omega]$ and the Clifford+T gate library is that the quantum operations which can be realized exactly by Clifford+T gates (i.e., without any rounding error) are precisely given by those matrices whose entries are from the ring $\mathbb{D}[\omega] = \mathbb{D}[\sqrt{2}, i]$ (as shown in [53]). As a consequence, all such quantum operations are represented with perfect accuracy. Hence, $\mathbb{D}[\omega]$ provides the ideal basis for a decision diagram that employs an accurate, algebraic representation of complex numbers.

---

[2] If it was, then also $\frac{1}{2}$ would be a member and could be written as $\frac{1}{2} = a' + b'\sqrt{2}$ for some $a', b' \in \mathbb{Z}$. However, since it must hold that $b' \neq 0$, this immediately yields the contradiction $\sqrt{2} = \frac{1-2a'}{2b'} \in \mathbb{Q}$.

[3] The fact that the rings $\mathbb{Z}[i, \frac{1}{\sqrt{2}}]$ and $\mathbb{D}[\omega]$ are isomorphic becomes obvious if one considers the ring $\mathbb{D}[\sqrt{2}, i]$ (which can be easily seen to be isomorphic to both rings) as an intermediate step. In fact, $\sqrt{2} = \omega - \omega^3$ and $i = \omega^2$.

8.2.2 Incorporating $\mathbb{D}[\omega]$ into DDs

In order to use the algebraic representation of complex numbers presented above within DDs as used in this part of the book, there are two aspects that have to be taken into account:

1. In order to determine common factors and structural similarities (that are required to find redundancies), a unique representation of $\mathbb{D}[\omega]$ numbers is required. However, there are in general infinitely many possibilities to represent a $\mathbb{D}[\omega]$ number.
2. The extracted (normalization factors) have to be applied to the edge weights. More precisely, the weights have to be divided by these factors. However, as division means multiplication by the (multiplicative) inverse, this division can only be conducted properly for $\mathbb{D}[\omega]$ numbers that indeed have a multiplicative inverse in $\mathbb{D}[\omega]$, but not for $\mathbb{D}[\omega]$ numbers in general (e.g., all odd integers ≥ 3 do not have an inverse in $\mathbb{D}[\omega]$ and the result of a division by such a number cannot be represented as a $\mathbb{D}[\omega]$ number).

These issues are addressed as follows:

1. Recall that each number from $\mathbb{D}[\omega]$ can be written as

$$\alpha = \frac{1}{\sqrt{2}^k}(a\omega^3 + b\omega^2 + c\omega + d)$$

for coefficients $a, b, c, d, k \in \mathbb{Z}$. If the exponent k is fixed, the representation is clearly unique since two different representations would yield a non-trivial representation of 0 in $\mathbb{Z}[\omega]$.[4] Thus, a unique representation can be achieved when using the *smallest denominator exponent* k_{min} such that there is no representation with an exponent $k < k_{min}$.

The existence of such an exponent has already been discussed in [53], but no constructive criterion for minimality has been derived. To this end, note that $\sqrt{2} = -\omega^3 + \omega$, such that

$$\alpha = \frac{1}{\sqrt{2}^k}(a\omega^3 + b\omega^2 + c\omega + d) \cdot \frac{\sqrt{2}}{\sqrt{2}}$$

$$= \frac{1}{\sqrt{2}^{k+1}}\left((b-d)\omega^3 + (c+a)\omega^2 + (b+d)\omega + (c-a)\right)$$

$$= \frac{1}{\sqrt{2}^{k-1}}\left(a'\omega^3 + b'\omega^2 + c'\omega + d'\right)$$

---

[4]This would contradict the fact that the potencies $\omega^0 = 1, \omega, \omega^2, \omega^3$ are linearly independent over $\mathbb{Z}$ (even over $\mathbb{Q}$), since ω is a primitive 8-th root of unity and the cyclotomic field $\mathbb{Q}[\omega]$ is a 3-dimensional vector space over $\mathbb{Q}$.

where $a', b'c', d' \in \mathbb{Z}$ if, and only if, $a = c \mod 2$ and $b = d \mod 2$. Thus, the exponent is known to be minimal if, and only if, $a \neq c \mod 2$ or $b \neq d \mod 2$.[5]

In summary, the above consideration yields a constructive algorithm to obtain unique representations of $\mathbb{D}[\omega]$ numbers.

2. The fact that a similar argumentation as above can be performed for the field $\mathbb{Q}[\omega]$ as well is exploited in the division by normalization factors. In fact, each $\mathbb{Q}[\omega]$ number has a unique representation as $\frac{\alpha}{e}$ where $\alpha \in \mathbb{D}[\omega]$ and e is an odd integer ($e \in 2\mathbb{Z} + 1$) that is co-prime to the integer coefficients of α.

Having this, all computations can be made in the field $\mathbb{Q}[\omega]$ where all non-zero numbers have a multiplicative inverse. This inverse can be constructed as follows: The squared norm $N(z)$ of a number $z \in \mathbb{Q}[\omega]$ is given as

$$N(z) = zz^* = u + v\sqrt{2} \text{ for some } u, v \in \mathbb{Q}.$$

Using the third binomial formula, the inverse of $N(z)$ can, hence, be written as

$$\frac{1}{N(z)} = \frac{u - v\sqrt{2}}{u^2 - 2 \cdot v^2}.$$

Finally, the inverse of z is given by rewriting the first equation as

$$z^{-1} = z^* \cdot \frac{1}{N(z)}.$$

In summary, by spending one additional integer and switching to the algebraic number field $\mathbb{Q}[\omega]$, a division/normalization becomes possible.

Overall, this allows for the algebraic and, thus, perfectly accurate representation of complex numbers within the DD representation presented in Chap. 5.

8.3 Evaluation

This section presents the results of the conducted evaluation. More precisely, a detailed evaluation is conducted on the current trade-off between accuracy and compactness in decision diagrams for quantum computations following the *numerical DD representation* discussed above (which utilizes floating point numbers in the IEEE 754 double precision format to represent irrational coefficients and supports the configuration of a tolerance value ϵ). Note that this evaluation only becomes possible by having a perfectly accurate solution, namely the presented *algebraic DD*

---

[5]Since there is no smallest denominator exponent for 0, the unique representation of 0 is defined as $a = b = c = d = k = 0$.

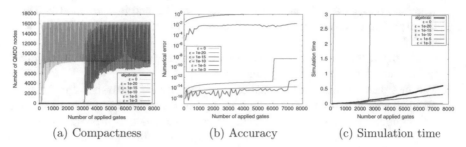

(a) Compactness   (b) Accuracy   (c) Simulation time

Fig. 8.2 Results for simulating Grover's algorithm

representation, to compare with. In a second step, it is evaluated how the presented solution overcomes this trade-off.

To this end, several well-known quantum algorithms available (including Grover's algorithm for database search [59], the Binary Welded Tree (BWT, [26]) algorithm, as well as the GSE algorithm already used in Example 8.2) are simulated with the DD-based method described in Chap. 5 on a 3.8 GHz machine with 32 GB of memory. In the following, a summary of the respectively obtained results for a representative selection of benchmarks is provided and discussed.

8.3.1   Trade-Off Between Accuracy and Compactness

The first part of the evaluation investigated the accuracy and the compactness (as well as the impact on the simulation run-time) of the recently applied, i.e., *numerical*, DD representation for different values of ϵ. In the following, the obtained results are discussed for Grover's algorithm as well as for the BWT and GSE algorithm whose results are provided in Figs. 8.2, 8.3, and 8.4, respectively, and which provide good representatives of the conducted evaluation. For each quantum algorithm, the graphs represent the size of the DD (i.e., the number of DD-nodes), the accuracy throughout the simulation,[6] as well as the run-time of the simulation (in CPU seconds).

First, the results clearly confirm the general numerical stability of the DD-based matrix multiplication (cf. Sect. 8.1). In fact, for a sufficiently small tolerance value ϵ, the error indeed scales linearly with the number of applied gates. In addition, the provided plots also show that the compactness of the DD directly correlates with

---

[6]The accuracy is quantified as the relative deviation of the vector resulting from the numerical computation v_{num} from the algebraic (and, thus, exact) result v_{alg}. More precisely, the Euclidean norm of $v_{num} - v_{alg}$ quantifies the loss of precision relative to the exact result. To have a fair evaluation, the norm of the numerically computed vector v_{num} is adjusted to 1, since an error in the length of the vector can be easily fixed (except for a 0-vector).

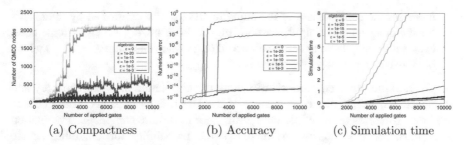

(a) Compactness   (b) Accuracy   (c) Simulation time

Fig. 8.3 Results for simulating the BWT algorithm

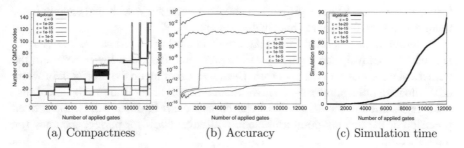

(a) Compactness   (b) Accuracy   (c) Simulation time

Fig. 8.4 Results for simulating the GSE algorithm

the simulation time. More precisely, the slope of the simulation times depicted in Figs. 8.2c, 8.3c, and 8.4c is proportional to the respective number of DD-nodes.

However, the results also clearly confirm the discussed trade-off between accuracy and compactness: Consider, for example, the plots obtained for simulating Grover's algorithm using 15 qubits (i.e., Fig. 8.2):

- Using a numeric DD representation with a high accuracy ($\epsilon = 0$ or $\epsilon = 10^{-20}$) hardly allows detecting any redundancies and, thus, requires exponentially many nodes and a significant run-time.
- In contrast, choosing a moderate accuracy ($\epsilon = 10^{-15}$ or $\epsilon = 10^{-10}$) allows detecting more of the actually present redundancies and, hence, yields a quite compact representation. On the downside, this truncation leads to numerical issues. For instance, while choosing $\epsilon = 10^{-15}$ yields a rather small numerical error, the peaks in the graph indicate an undesired numerical instability in the multiplication algorithm that may lead to severe rounding errors in certain simulations.
- By choosing a low accuracy ($\epsilon = 10^{-5}$ or $\epsilon = 10^{-3}$) the accuracy of the DD-based representation drops significantly—resulting in completely useless simulation results. Surprisingly, the number of DD-nodes even increases exponentially for $\epsilon = 10^{-5}$ after applying approximately 3000 gates. However, this is an exceptional case since in the vast majority of the cases, increasing ϵ indeed

increases the compactness of the numeric DD (as also confirmed by the results for the other quantum algorithms, cf. Figs. 8.3a and 8.4a). As an extreme case, even a dropping down to zero is observed (e.g., when choosing $\epsilon = 10^{-3}$) which obviously is a completely wrong result.

Overall, the provided plots clearly show the correlation between accuracy and compactness. In fact, the compactness of the DD heavily depends on the chosen accuracy (in terms of ϵ). While increasing ϵ yields a more compact representation and, thus, reduces run-time, it increases the probability for obtaining severe numerical errors—resulting in completely useless results (e.g., a zero-vector) in the worst case. As shown above, a good choice of ϵ depends on the considered problem instance and it can be quite difficult to choose a value for ϵ such that exactly those redundancies are found that are actually present.

In addition, the plots also show that, even when using a tolerance value of $\epsilon = 0$, i.e., when employing the highest possible precision, there is a lower bound to the numerical error that is never underrun. Even when scaling up the precision/bitwidth of the floating point numbers—an investment that will likely lead to substantial run-time degradations—the same effect can be expected. In other words, the limited precision of the floating-point arithmetic will never allow for perfect accuracy (on the long run).

8.3.2   Evaluation of the Algebraic Representation

The algebraic DD solution presented in this chapter overcomes the limitations that have been observed in the previous evaluation. In fact, there is no more need for determining an adequate accuracy for the problem at hand on a case-by-case basis. The algebraic DD will always achieve the maximum compactness that is possible without losing information (i.e., only exploiting redundancies that are actually present). Moreover, it will achieve perfect accuracy also on the long run, which can be very important for design tasks like verification. For instance, checking equivalence of two matrices or vectors then boils down to comparing the root nodes of the corresponding DDs (which can be done in $O(1)$) instead of looking for (tiny) deviations in the whole representations.

However, the algebraic representation of complex numbers requires performing arithmetic operations in the ring $\mathbb{Q}[\omega]$. These can induce a computation overhead as the integer coefficients may (in theory) become arbitrary large, while floating-point arithmetic can often benefit from existing hardware accelerators, e.g., in terms of a dedicated floating point unit. In the following, this computation overhead is evaluated in detail.

To this end, Grover's as well as the BWT and GSE quantum algorithm as discussed above again provide a good representative of the conducted evaluation (also including an example with a worst case overhead). Accordingly, corresponding algebraic DDs are generated and the respectively obtained DD sizes and run-times by means of the bold black graphs are reported in Figs. 8.2, 8.3, and 8.4.[7]

As can be seen, for Grover's as well as for the BWT algorithm, the algebraic DDs remain quite compact. Thus, they perform much better than the numerical DDs with high accuracy ($\epsilon = 0$ and $\epsilon = 10^{-20}$) that cannot take advantage of the present redundancies. In comparison to the numerical DDs that exploit these redundancies, the algebraic DDs have a reasonable constant run-time overhead (around a factor of 2).

In contrast, the GSE algorithm is a representative for those cases, where the behavior is quite different. As can be seen, there are hardly any redundancies that can be exploited such that the size of the algebraic DD stays in the range of the sizes of the highly accurate numerical DDs. However, unlike in the previous cases, the run-times of the algebraic DD do not stay in the range of those numerical DDs that show the same sizes. In fact, they do not stay even close to the run-times of *any* numerical DD—the computation overhead grows significantly. A more detailed analysis shows that this is explained by the fact that the bitwidths of the integers used for algebraically representing the occurring complex numbers (especially for the denominator of the $\mathbb{Q}[\omega]$ numbers) grow significantly. This issue is related to the way the DD is normalized and might be resolved by alternative normalization schemes that keep the edge weights in $\mathbb{D}[\omega]$.

Overall, in several cases the structural benefits of algebraic DDs in comparison to numerical DDs really become effective as the computation overhead remains small, while in other cases a significant overhead is observed, which might turn out to be the showstopper for algebraic DDs in such cases. Nevertheless, algebraic DDs instead of numeric ones might significantly improve state-of-the-art approaches for synthesis [116, 117, 119], verification [118, 162, 167, 181], and simulation (cf. Chap. 5).

---

[7]Note that no graph is provided for the accuracy of the presented algebraic representation since it is always exact, i.e., does not include any numerical errors.

Chapter 9
Summary

Abstract This chapter summarizes Part II of this book by reflecting the capabilities of the proposed DD-based simulation approach, as well as the covered improvements compared to the state of the art.

Keywords Quantum circuits · DD-based simulation · Compact representation

This part of the book introduced the foundations of a DD-based simulation approach as well as a corresponding efficient implementation of the underlying DD-package that allows significantly outperforming straightforward Schrödinger-style simulators as well as previous DD-based simulators in many cases—even though these techniques have been heavily optimized over the last decade and utilize multiple CPU-cores to reduce simulation time (while the presented DD-based approach utilizes a single core only). More precisely, the presented approach is capable of (1) simulating quantum computations for more qubits than before, (2) in significantly less run-time (in hours or, in many cases, just minutes or seconds rather than several days), and (3) on a regular Desktop machine.

Even though the initial implementation of the presented DD-based simulation approach reached a substantial improvement compared to previous simulators, this new method is pushed further to its limit by exploiting further potential of DDs in the context of quantum-circuit simulation. More precisely, matrix–matrix multiplication is not necessarily more expensive than matrix–vector multiplication when using DDs for their representation (which is contrary to simulation approaches using array-based representations). Exploiting this observation allows developing strategies for combining operations before applying them to the state vector—leading to further speedups of several factors or, when additionally exploiting further knowledge, even of several orders of magnitude compared to the initial implementation.

Enormous improvements compared to the state of previous simulators obviously require an efficient implementation of the underlying DD-package—especially for handling the occurring complex numbers. By providing such techniques—in joint consideration of implementation techniques for decision diagrams in the

© Springer Nature Switzerland AG 2020   89
A. Zulehner, R. Wille, *Introducing Design Automation for Quantum Computing*,
https://doi.org/10.1007/978-3-030-41753-6_9

conventional domain developed decades ago—the development of a fully fletched DD-package for the quantum domain is leveraged. The conducted evaluation of the developed DD-package showed that it is indeed capable of handling complex numbers much more efficiently than previous implementations and allows constructing decision diagrams for established quantum functionality in significantly less runtime (up to several orders of magnitude). Presumably, this performance boost can be easily passed to DD-based methods for other design automation tasks like synthesis [116, 117, 119] or verification [118, 162, 167, 181], just by incorporating this new package.

Since handling complex numbers is crucial in DDs for quantum computation (especially when occurring as edge weights), the resulting trade-off between accuracy and compactness has been thoroughly discussed and evaluated. Since this trade-off requires fine-tuning of parameters on a case-by-case basis and might still yield useless results, an algebraic decision diagram is presented that overcomes this issue. The presented algebraic representation guarantees perfect accuracy while remaining compact (all redundancies that are actually present are detected). The conducted evaluation confirms the trade-off in the numerical representations and shows that the overhead of the algebraic solution is moderate in many cases.

All the endeavors listed above have been implemented in C/C++ and made publicly available at http://iic.jku.at/eda/research/quantum_simulation. Besides that, a stand-alone version of the developed DD-package is available at http://iic.jku.at/eda/research/quantum_dd. Together with the significant improvements gained compared to the state of the art, this did not only result in acknowledgment inside the academic community through publications at top-notch venues, but also received interest from big players in the field. More precisely, the developed simulation approach has been acknowledged with a *Google Research Faculty Award* and has recently been officially integrated into IBM's SDK *Qiskit* and Atos QLM. This further emphasizes the potential of DD-based design methods in the quantum domain—hopefully leading to as powerful DD-based methods as taken for granted in the conventional domain today.

Part III
Design of Boolean Components for Quantum Circuits

Chapter 10
Overview

Abstract This chapter set the context of reversible-circuit synthesis methods (required to design Boolean components occurring in quantum algorithms) presented in this part of the book. To this end, existing synthesis approaches are discussed, including their benefits and drawbacks. Eventually, cost metrics for the synthesized circuits are presented, which are utilized in the following chapters to quantify and measure the performance for the developed algorithms.

Keywords Quantum circuits · Boolean components · Boolean functions · Synthesis · Reversible circuits · Qubits · embedding

Estimating resource requirements of quantum algorithms (i.e., the number of required qubits and run-time on quantum computers), their simulation, or their execution on real hardware requires compiling quantum algorithms containing high-level operations (e.g., modular exponentiation in Shor's algorithm) into quantum circuits composed of elementary gates available on the considered target architecture. Thereby, quantum circuits composed of multiple-controlled one-qubit gates are usually considered since they (1) describe a rather low-level but still technology independent description of the algorithm, (2) can be directly handled by most simulators (cf. Sect. 5.2.1), and (3) are usually utilized as input for technology mapping algorithms (which will be covered in Part IV of this book).

For the "quantum part" of an algorithm, a decomposition into multiple-controlled one-qubit gates is usually inherently given by the algorithm, by using common building blocks like a *Quantum Fourier Transform* (QFT [45]), or determined by hand (even though few automated approaches exist [116, 117, 119]). However, this is different for large Boolean components that are contained in many quantum algorithms, e.g., the modular exponentiation in Shor's algorithm for integer fac-

© Springer Nature Switzerland AG 2020   93
A. Zulehner, R. Wille, *Introducing Design Automation for Quantum Computing*,
https://doi.org/10.1007/978-3-030-41753-6_10

torization [143] or a Boolean description of the database that is queried in Grover's algorithm [59].[1]

Even though the functionality of the Boolean components can be described in the conventional domain, corresponding design methods cannot be utilized since the inherent reversibility of quantum computations has to be considered. In fact, determining circuits composed of reversible gates only requires dedicated *reversible-circuit synthesis* approaches [150, 151]. This becomes crucial since the Boolean components commonly describe very complex functionality and, thus, are split into several (non-)reversible parts to become manageable [13, 62, 63, 147, 152]. However, these resulting non-reversible sub-functions have to be *embedded* into reversible ones to ensure the desired unique mapping from inputs to outputs—a task that can either be conducted explicitly or implicitly (e.g., by certain synthesis methods discussed below). Since this embedding process requires adding several so-called *ancillary qubits*, how to efficiently realize these non-reversible Boolean sub-functions in a reversible fashion while keeping the number of required qubits as low as possible (since they are a highly limited resource) has received significant interest, and is prioritized in the following.

T-count and T-depth (cf. Sect. 2.3) of the synthesized reversible circuits serve as cost metric to compare different approaches that yield circuits with an equal (or at least an almost equal) number of qubits.[2] This is motivated by the fact that quantum algorithms containing large Boolean components are envisioned to run on future fault-tolerant quantum computers (rather than on NISQ devices).

In the past, two directions (which are summarized in Fig. 10.1) emerged that address these objectives in a different fashion.

Structural Synthesis (sketched at the bottom of Fig. 10.1) used conventional methods such as *Binary Decision Diagrams* (BDDs, [171]), *Exclusive-Or Sum-of-Products* (ESoP, [48]), gate netlists [184], or *Lookup Tables* (LUTs, [152]) to synthesize the function in the conventional domain first. Then, each corresponding building block such as a BDD-node, a product/exclusive sum, a primitive gate, or a lookup table is mapped to a functionally equivalent cascade of quantum gates. As all these building blocks are non-reversible, the equivalent cascade of gates usually requires additional qubits. Since numerous such building blocks are mapped for larger functions, this leads to a number of working qubits which may be orders of magnitude larger than the actual minimum (as, e.g., evaluated in [174])—even though if post-synthesis optimization (e.g., [178, 179]) and adjusted synthesis schemes (e.g., [147]) are utilized.

As an alternative, *Functional Synthesis* (sketched at the top of Fig. 10.1) has been proposed. Here, a non-reversible function is embedded into a reversible one

---

[1]The general ideas of these algorithms are that the Boolean component is evaluated with a highly superposed input (e.g., by applying Hadamard gates first) to gain speedup through quantum parallelism.

[2]Since there exists a trade-off between the number of ancillary qubits as well as the T-count and T-depth of a circuit [179], comparing T-count and T-depth of circuits with a significantly different number of qubits is meaningless.

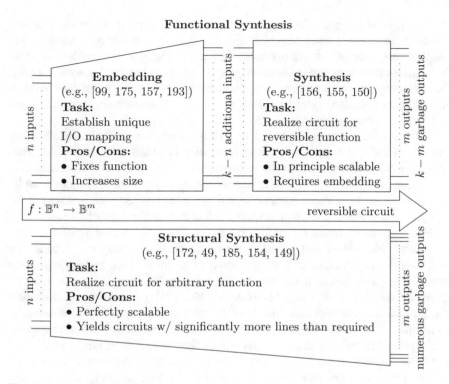

Fig. 10.1 Reversible circuit synthesis

prior to synthesis. This *embedding* step [97, 155, 174, 192] adds further variables to the function in order to distinguish non-unique output patterns, while keeping the overhead minimal (in contrast to structural approaches). Afterwards, the function is passed to the *actual synthesis* method, which eventually yields a quantum circuit. Corresponding synthesis approaches for reversible circuits range from exact solutions [58] to heuristic ones, e.g., based on truth tables [103, 140], positive polarity Reed–Muller expansion [61], or Reed–Muller spectra [98]. However, since all these approaches rely on an exponential description of the underlying function, they are limited to rather small functions. In order to improve this limited scalability, alternative synthesis approaches have recently been proposed that explicitly exploit efficient data-structures such as decision diagrams [153, 154] or are based on Boolean satisfiability [148]. Nevertheless, there is still room for improvement with respect to scalability and costs of the synthesized circuit—enabling to synthesize larger Boolean components with minimal overhead. Besides that, a significant degree of freedom is neglected by the current functional design flow by fixing the embedding explicitly before synthesis—even though there might exist other reversible embeddings (requiring the same number of qubits) that are more suitable for the synthesis approach and, hence, result in a circuit with lower costs.

This part of the book focuses on the *functional* design flow for synthesizing Boolean components since structural approaches yield circuits with a substantial number of qubits (which are a limited resource). Thereby, the problem is again investigated from a design automation perspective to exploit (yet) unused potential that allows synthesizing cheaper circuits, yields better scalability, and even reduces the number of required qubits below what is currently considered as the minimum (under certain assumptions).

More precisely, Chap. 12 (based on [192]) presents a scalable methodology (based on DDs) for embedding non-reversible functions into reversible ones. Moreover, several improvements for an already existing DD-based synthesis method are presented (based on [191]) that often reduce the costs of the resulting circuit by orders of magnitude. These methods significantly contribute to the progress in the design of reversible circuits with respect to efficiency and scalability.

However, this currently established design flow still suffers from the need to conduct embedding and actual synthesis separately—a major drawback that prohibits the exploitation of a huge degree of freedom since embedding is not necessarily conducted in a fashion, which suits the following synthesis step. To overcome this drawback, Chap. 13 (based on [193, 196]) presents a completely new design flow that combines functional synthesis and the embedding to a *one-pass design flow*. This generic flow is not bound to a certain functional synthesis approach and exploits the available degree of freedom to significantly increase scalability and to reduce the costs of the synthesized circuit while keeping the number of required qubits at the minimum. Moreover, it is also shown (based on [188, 195, 197]) that the design flows can be enriched with coding techniques to reduce the number of qubits that have to be considered during synthesis. In fact, a single ancillary qubit is sufficient to realize any desired non-reversible function in a reversible fashion. Assuming that no decoding is required (e.g., when the subsequent Boolean component is capable of handling encoded inputs) this allows pushing the number of required qubits far below what is usually considered as minimum. Even if decoding is necessary (again increasing the number of required qubits to the minimum), the considered coding strategy turns out to be beneficial since it significantly reduces the costs of the resulting circuit.

Finally, the findings discussed in this part of the book are summarized in Chap. 14. In order to realize all these endeavors in an efficient fashion, the key idea is to express the required tasks as matrix operations, which are conducted efficiently by using compact representations for matrices similar to the decision diagrams introduced in Chap. 5. To this end, Chap. 11 (based on [121]) first discusses how such DDs are utilized to represent Boolean functions, and how they are constructed without building the exponentially large matrix at first hand. Note that in the following matrices and DDs are used interchangeably to describe the presented methods, since the decision diagram can be seen as black box providing matrix operations—allowing to describe the presented methods from a matrix perspective whenever this is more suitable.

Chapter 11
Efficient Representation of Boolean Components

Abstract This chapter discusses efficient representations for Boolean components occurring in quantum algorithms. More precisely, Boolean functions are first represented as matrices, where columns and rows represent input and output patterns, respectively. Then, decision diagrams (similar to those of the DD-based simulator presented earlier in this book) are utilized in order to gain a compact representation of these matrices. The algorithms of the remaining chapters of this part of the book then utilize matrix operations to solve the considered synthesis tasks.

Keywords Boolean functions · Permutation matrix · Function matrix · Decision diagrams · BDD

Even though the functionality of the Boolean components occurring in quantum circuits can be described in the conventional domain, the inherent reversibility of quantum computing as well as its realization as reversible circuit must be taken into account. Thereby, a compact representation serves as basis for efficient reversible-circuit synthesis methods. Since the complex functionality of Boolean components is usually decomposed in several non-reversible sub-function, corresponding description means are also required.

To satisfy these requirements, this chapter (based on [121]) describes Boolean reversible functions by means of *permutation matrices* (of size $2^n \times 2^n$), which are compactly represented using the DDs developed for quantum-circuit simulation (cf. Sect. 5.1). In order to support non-reversible functions also, the so-called function matrices are introduced, which again are compactly represented using DDs. Moreover, this chapter (based on [121]) provides methods to generate these DDs directly from a compact function description like BDDs and without generating the exponentially large matrix at first hand.

In the following, the utilized DD-based representation is presented. Section 11.1 discusses the description of Boolean functions as function matrices, which then are compactly represented using DDs. Section 11.2 presents methods to determine these DDs without the need of generating the exponentially large matrix.

11.1 Decision Diagrams for Boolean Components

Since quantum computing is inherently reversible, the occurring Boolean components describe a permutation of the input combinations that, eventually, are represented by so-called a *permutation matrix*.

Definition 11.1 *Let $f : \mathbb{B}^n \to \mathbb{B}^n$ be a reversible function. A permutation matrix M representing f is a matrix of dimension $2^n \times 2^n$ in which each column (row) of the matrix represents one possible input pattern (output pattern) of f. The elements $m_{i,j}$, $0 \le i, j < 2^n$ of the matrix M are defined by*

$$m_{i,j} = \begin{cases} 1 & \textit{if } f(j) = i, \\ 0 & \textit{otherwise.} \end{cases}$$

The columns (rows) of a permutation matrix represent the inputs (outputs). If an input maps to an output, the corresponding entry of the matrix is set to 1. All other entries in the permutation matrix are set to 0. Note that each row and each column of a permutation matrix (representing a reversible function) contains exactly one 1-entry (caused by the one-to-one mapping).

Example 11.1 *Consider the reversible function provided in terms of a truth table in Fig. 11.1a. The functionally equivalent permutation matrix is provided in Fig. 11.1b.*

Since a permutation matrix is a special case of a unitary matrix (where all entries are either 0 or 1), the DD-based representation introduced in the context of quantum-circuit simulation (cf. Sect. 5.1) can be utilized to represent them efficiently. In fact, the representation of permutation matrices is even simpler, since all edge weights are guaranteed to be either 1 or 0.

Example 11.1 (continued) *Figure 11.1c shows a DD-based representation of the permutation matrix shown in Fig. 11.1b.*

As discussed in Chap. 10, Boolean reversible functions are often broken down into non-reversible ones that have to be realized in a reversible fashion, i.e., with a reversible circuit. In order to have a common description mean for reversible and

x_0	x_1	y_0	y_1
0	0	1	1
0	1	1	0
1	0	0	0
1	1	0	1

(a) Truth table

Inputs

	00	01	10	11
00	0	0	1	0
01	0	0	0	1
10	0	1	0	0
11	1	0	0	0

Outputs

(b) Permutation matrix

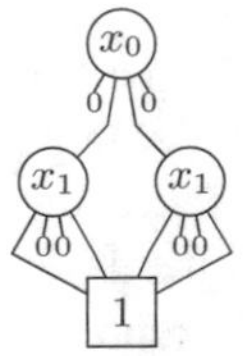

(c) DD-based representation

Fig. 11.1 Representations for a reversible function

non-reversible functions, permutation matrices are generalized in a straightforward fashion to *function matrices* that—similar to permutation matrices—strictly tie each input to an output.

Definition 11.2 *Let $f : \mathbb{B}^n \to \mathbb{B}^m$ be a Boolean function. Then, the function matrix M of f is a $2^k \times 2^k$ matrix with $k = \max(n, m)$ and elements $m_{i,j}$, $0 \leq i, j < 2^k$ such that*

$$m_{i,j} = \begin{cases} 1 & \text{if } f(j) = i, \\ 0 & \text{otherwise.} \end{cases}$$

Again, the rows and columns represent the outputs and input, respectively. Consequently, each column of a function matrix contains at most one 1-entry. In contrast to permutation matrices, a row may contain multiple 1-entries, because more than one input combination may map to the same output pattern. Nevertheless, DDs as introduced in Sect. 5.1 are capable for representing function matrices since they have a dimension of $2^k \times 2^k$.

Example 11.2 *Figure 11.2a shows the truth table of a half adder. The corresponding function matrix representation is depicted in Fig. 11.2b, while the corresponding DD is depicted in Fig. 11.2c. For example, the fourth column of the function matrix represents the input 11, which is supposed to map to the output 10. Hence, the fourth column (representing input 11) contains its 1-entry in the third row (representing output 10).*

This discussed compact DD-based representation of function matrices allows for directly applying matrix operations like addition and multiplication (cf. Chap. 5) and, thus, is perfectly suited for the design tasks considered in this part of the book. However, in most cases the desired functionality is originally not provided in terms of such a DD, but using alternative representations such as Boolean algebra or circuit netlists. Hence, the DD representing the considered functionality has to be constructed before applying the corresponding approaches. How this can be done—without building an exponentially large intermediate matrix at first hand—is discussed in the next section.

x_0	x_1	y_0	y_1
0	0	0	0
0	1	0	1
1	0	0	1
1	1	1	0

(a) Truth table

		00	01	10	11
Outputs	00	1	0	0	0
	01	0	1	1	0
	10	0	0	0	1
	11	0	0	0	0

Inputs

(b) Function matrix

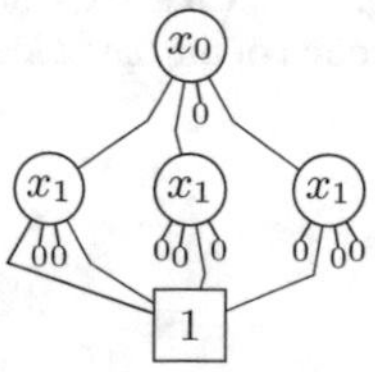

(c) DD representation

Fig. 11.2 Representations for the half adder function

11.2 Efficient Construction of DDs for Function Matrices

This section shows how a DD representing function matrix M of a Boolean function f is directly constructed from a compact representation of f (i.e., by means of BDDs; cf. Sect. 3.1). In order to bridge the gap between an initial representation and the targeted DD representation, the main idea is to employ the so-called *characteristic function* χ_f of f. This is a Boolean function $\mathbb{B}^n \times \mathbb{B}^m \to \mathbb{B}$ with n inputs labeled $x = x_0, \ldots, x_{n-1}$ and m inputs labeled $y = y_0, \ldots, y_{m-1}$, where $\chi_f(x, y) = 1$ if, and only if, $f(x) = y$. In other words, χ_f evaluates to true if, and only if, the backmost m inputs represent the correct output pattern that is generated when applying f to the input pattern specified by the first n inputs. Thus, the entries of the function matrix M (when traversing the matrix row-wise) can be interpreted as the outcomes of χ_f.

Example 11.3 *The characteristic function of the half adder from Example 11.2 is shown in Fig. 11.3 in terms of its truth table. Each line corresponds to one entry of the function matrix. More precisely, writing all columns of the function matrix on top of each other would yield the χ_f column of the truth table.*

As it is infeasible to construct and store the whole function matrix at once (due to its exponential complexity), compact, graphical representations of Boolean functions (especially of the characteristic functions) are employed instead, from which the desired DD representation is then derived directly without explicitly considering the function matrix. To this end, BDDs (cf. Sect. 3.1) are utilized.

Example 11.4 *The BDDs for the outputs of the half adder reviewed in Examples 11.2 and 11.3 are shown on the left-hand side of Fig. 11.4.*

Overall, there is a well-developed methodology for constructing the BDD representation of the outputs of f. These BDDs have then to be composed in a second step to obtain the BDD of the characteristic function χ_f. Since the outcomes of χ_f essentially describe the entries of the desired function matrix, the resulting BDD can eventually be transformed to a DD as introduced in Sect. 5.1.2. In the following, these steps are described in more detail.

Fig. 11.3 Characteristic function of the half adder

x_0	x_1	y_0	y_1	χ_f
0	0	0	0	1
0	0	0	1	0
0	0	1	0	0
0	0	1	1	0
0	1	0	0	0
0	1	0	1	1
$\vdots$	$\vdots$	$\vdots$	$\vdots$	$\vdots$
1	1	1	1	0

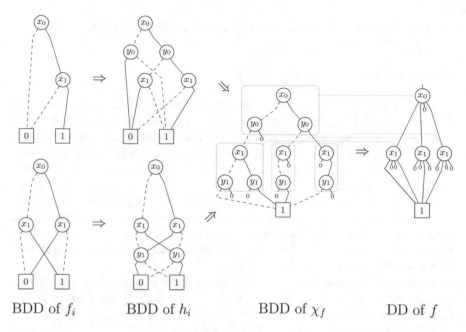

BDD of f_i BDD of h_i BDD of χ_f DD of f

Fig. 11.4 Construction of the DD for the half adder

11.2.1   Generating the BBD of the Characteristic Function

In order to derive the BDD representing the characteristic function χ_f of a multi output function $f : \mathbb{B}^n \to \mathbb{B}^m$, new variables y_i for the primary outputs of f (referred to as *output variables* in the following) are introduced. While the original (input) variables are used to encode the column index of the function matrix, the output variables encode rows. Then, the characteristic function is constructed for each output. More precisely, the helper functions h_i given by

$$h_i(x_0, \ldots, x_{n-i}, y_i) = f_i(x_0, \ldots, x_{n-1}) \odot y_i,$$

are constructed where $\odot$ denotes the XNOR-operation. This logical operation—and, thus, the entire function h_i—evaluates to true if, and only if, both operands are equal, i.e., if $f_i(x_0, \ldots, x_{n-1}) = y_i$. Consequently, the h_i-function can be interpreted as characteristic functions of the primary outputs of f.

Afterwards, the BDD of χ_f can be constructed by AND-ing the BDDs representing the h_i-functions as the following calculation shows:

$h_0 \wedge h_1 \wedge \ldots \wedge h_{m-1} = 1$

$\Leftrightarrow \forall i \in \{0, \ldots, n-1\} : h_i = 1$

$\Leftrightarrow \forall i \in \{0, \ldots, n-1\}, (x_0, \ldots, x_{n-1}, y_0, \ldots, y_{m-1}) \in \mathbb{B}^{n+m} : f_i(x_0, \ldots, x_{n-1}) = y_i$

$\Leftrightarrow f(x_0, \ldots, x_{n-1}) = (y_0, \ldots, y_{m-1})$

$\Leftrightarrow \chi_f(x_0, \ldots, x_{n-1}, y_0, \ldots, y_{m-1}) = 1.$

Note that if $n > m$, i.e., if f has more primary inputs than outputs, the function is padded with zeros in order to obtain a Boolean function with the same number of inputs and outputs, such that the resulting function matrix is square (cf. Definition 11.2). More precisely, $n - m$ additional constant outputs $f_j \equiv 0$ are added. While these can, in principle, be added at any position, they are added in front of the original outputs. If, in contrast, $m > n$, $m - n$ additional inputs are added, which have no impact on the functionality of f. Again, these inputs can, in principle, be added at any position, but they are added in front of the original inputs. Overall, this ensures that the original functionality is represented by the submatrix of dimension $2^m \times 2^n$ in the top-left corner of the square function matrix of dimension $2^k \times 2^k$, where $k = \max(n, m)$.

As the BDD representing χ_f is guaranteed to be exponential in size for the variable order $x_0 \succ \ldots \succ x_{k-1} \succ y_0 \succ \ldots \succ y_{k-1}$ (at least for reversible functions), an interleaved variable order $x_0 \succ y_0 \succ x_1 \succ y_1 \succ \ldots \succ x_{k-1} \succ y_{k-1}$ is enforced when constructing the BDD for χ_f.

Example 11.5 *Consider again the half adder example. The BDDs representing the helper functions $h_0 = f_0 \odot y_0$ and $h_1 = f_1 \odot y_1$ are computed using the BDD equivalent of the logical XNOR operation and are shown in Fig. 11.4 (next to the BDDs representing f_0 and f_1). By AND-ing these BDDs, the BDD representing χ_f is obtained, which is shown in the center of Fig. 11.4. In this BDD, all edges pointing to the zero-terminal are indicated by stubs for the sake of a better readability and to emphasize the similarity to the targeted DD.*

11.2.2 Transforming the BDD into the Desired DD Representation

With a BDD in interleaved variable order representing χ_f, the matrix partitioning employed by DDs developed in Sect. 5.1.2 is already laid out implicitly. In fact, corresponding bits of the column and row indices are represented by different, but adjacent variables (x_i and y_i), while the DDs combine these in a single variable. Consequently, the BDD of χ_f can be transformed into the DD representing the function matrix of f using the general transformation rule shown in Fig. 11.5. However, there are two special cases that have to be treated separately:

- If an input variable x_i is skipped (more precisely: a node labeled by y_i is the child of a node not labeled by x_i), this implies the x_i node would be redundant, i.e., high and low edge point to the same node. This case can be easily handled by

Fig. 11.5 General
transformation rule for
characteristic BDDs

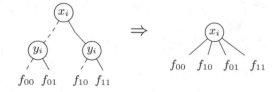

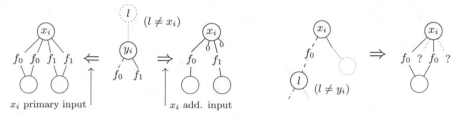

(a) Skipped input variables   (b) Skipped output variables

Fig. 11.6 Handling skipped variables

setting $f_{00} = f_{10} = f_0$ or $f_{01} = f_{11} = f_1$, respectively, as illustrated on the left-hand side of Fig. 11.6a. If, however, x_i is not an original input of the function, but has been introduced later in order to obtain the same number of in- and outputs, setting $f_{10} = f_{11} = 0$ instead to ensure that the original functionality occurs only once in the final function matrix (as illustrated on the right-hand side of Fig. 11.6a).

- If an output variable level y_i is skipped (more precisely: the high or low edge of a node labeled by x_i point to a node labeled by $l \neq y_i$), this implies the skipped y_i node would be redundant (both children would be the same). This case can be easily handled by setting $f_{00} = f_{01} = f_0$ or $f_{10} = f_{11} = f_1$, respectively, before applying the general transformation rule. For instance, the case of a skipped variable on the low edge is illustrated in Fig. 11.6b.

Example 11.6 *Consider again the characteristic BDD shown in the center of Fig. 11.4. Here, the single x_0 node and the leftmost x_1 node can be transformed to their function matrix DD equivalent by applying the general transformation rule. For the remaining x_1 nodes, the methodology for skipped y_1 output variables is applied. Overall, this yields the DD shown on the right-hand side of Fig. 11.4.*

Overall, following this procedure yields a DD representing the function matrix (in case of a reversible function, a permutation matrix) of any Boolean function f originally provided in terms of a BDD. By this, the efficient manipulation algorithms discussed in Sect. 5.1.2 are inherently available when constructing Boolean components occurring in quantum algorithms.

Chapter 12
Functional Synthesis

Abstract This chapter presents significant improvements for *functional synthesis* of reversible circuits by providing a new approach for embedding a non-reversible function into a reversible one while using the minimum number of ancillary qubits. Thereby, the decision diagrams developed for quantum-circuit simulation are utilized to compute the number of additionally required qubits in negligible run-time. Moreover, this chapter takes a closer look on an existing approach for synthesizing a quantum circuit given an embedded (i.e., a reversible) function, which is also based on decision diagrams, but can be significantly improved. More precisely, exploiting redundancies in nodes and paths of the decision diagram allows reducing the costs of the resulting circuit by several orders of magnitude in many cases.

Keywords Functional synthesis · Embedding · Qubits · Decision diagrams · DD nodes · Reversible circuits

This chapter investigates functional synthesis of reversible circuits (composed of an embedding and a synthesis step) while considering the efficient DD-based representation of Boolean functions discussed in Chap. 11. In fact, having a compact description means serves as basis for developing scalable embedding and synthesis approaches.

This chapter presents a functional synthesis approach for reversible circuit that is entirely based on function matrices (which are eventually represented by means of DDs). To this end, a new embedding approach following this description means is provided (based on [192]) that allows embedding larger function than the current state of the art. Moreover, the DD-based synthesis scheme originally proposed in [154] (which relies on permutation matrices) is revisited and significantly improved (based on [191]) by exploiting redundancies in paths and nodes of the underlying DD. This optimization reduces the T-count of the synthesized circuits by orders of magnitude in many cases.

The following sections present this functional synthesis approach (based on function matrices). Section 12.1 describes the considered embedding scheme.

© Springer Nature Switzerland AG 2020 105
A. Zulehner, R. Wille, *Introducing Design Automation for Quantum Computing*,
https://doi.org/10.1007/978-3-030-41753-6_12

Section 12.2 presents a synthesis method to realize the embedded function as reversible circuit while exploiting redundancies in the paths and nodes of the underlying DD.

12.1   Embedding

As stated in Chap. 10, functional approaches for reversible circuit synthesis require a reversible function as input. However, if the function to be synthesized is of non-reversible fashion (e.g., a sub-component of a Boolean oracle), a process called *embedding* [97] is applied, where the desired (non-reversible) function is enriched by additional inputs and outputs to facilitate reversibility. Since qubits are a limited resource, this overhead shall be kept at the minimum. Under this premise, determining an embedding for a non-reversible function has been proven to be a co$\mathcal{NP}$-hard problem in [155]—yielding an exponential complexity. Even just determining how many additional inputs and outputs are actually required in order to make a function reversible is very challenging.

Initial methods aiming for the determination of embeddings for non-reversible Boolean functions considered truth tables [97]—which obviously is not feasible for larger functions. Recently, more efficient approaches have been proposed, which rely on function representations such as binary decision diagrams [155]. But also here, the exponential complexity cannot really be tackled and, hence, results for functions with up to 30 variables only have been obtained thus far. Consequently, many large reversible functions are currently realized with a number of inputs/outputs that is orders of magnitude away from the actual minimum [174].

This section (based on [192]) tackles the underlying exponential complexity problem of embedding in a more scalable and entirely new fashion. More precisely, DDs as discussed in Chap. 11 are utilized to represent the function matrix of the function to be embedded. This matrix representation allows deriving the minimum number of additionally required outputs and, based on this number, determining a reversible embedding in a much more efficient fashion. The efficiency is gained through dividing the embedding process into matrix operations, which are directly employed on the DD. The conducted evaluation shows that this DD-based approach outperforms previous methods by several orders of magnitude in terms of run-time. Moreover, exact results on the number of additionally required outputs as well as on corresponding embeddings are obtained for some of the benchmarks for the first time ever.

In the following, the DD-based approach is described. Section 12.1.1 provides detailed overview of the embedding process and its main challenges. Then, a DD-based approach for determining the minimum number of additional outputs as well as for deriving a corresponding reversibly embedded function is described in Sect. 12.1.2. Section 12.1.3 presents the conducted evaluation of this approach as well as a comparison to the state of the art.

12.1.1 Existing Embedding Process

Consider Boolean functions $f : \mathbb{B}^n \rightarrow \mathbb{B}^m$ with n primary inputs and m primary outputs. If f is non-reversible, multiple input combinations are mapped to the same output pattern. Since reversibility requires a unique mapping from inputs to outputs, the output pattern must be made distinguishable. Therefore, additional outputs—so called garbage outputs—are added to the primary outputs.

Definition 12.1 *Let $f : \mathbb{B}^n \rightarrow \mathbb{B}^m$ be a Boolean function with n inputs and m outputs. Furthermore, let $\mu : \mathbb{B}^m \rightarrow \mathbb{N}$ be a function that provides the number of times a respective output pattern $p \in \mathbb{B}^m$ is generated by f. Then, the ith most frequent output pattern $(1 \leq i \leq 2^m)$ is denoted by p_i. In order to distinguish all occurrences of an output pattern p_i, at least $k_i = \lceil \log_2 \mu(p_i) \rceil$ garbage outputs are required. Since $k_i \leq k_1$ always holds (i.e., the number of garbage outputs required for an output pattern p_i is always less than or equal to the number of garbage outputs required for the most frequent output pattern p_1), the total number of garbage outputs required for a given function f is $k = k_1 = \lceil \log_2 \mu(p_1) \rceil$.*

Example 12.1 *Consider the non-reversible function as depicted in Table 12.1a. The function is non-reversible since 2 input combinations, namely 01 and 11, map to the same output pattern (01). Since 01 is the most frequent output pattern and occurs twice, $k = \lceil \log_2 2 \rceil = 1$ garbage output is required.*

Inserting garbage outputs results in extra columns of the truth table. Since the values of the garbage outputs are not of interest, they can be assigned arbitrarily. However, there are dependencies in the assignment: they have to be chosen in a way, such that the garbage outputs are assigned differently for all occurrences of an output pattern. In the following, this is represented by an asterisk (*).

Example 12.1 (Continued) *The garbage outputs of input patterns 01 and 11 depend on each other, because these inputs map to the same output. As soon as the garbage output for one of the input patterns is fixed to 1 (0), the garbage output for the other input pattern must be 0 (1) to ensure reversibility.*

Table 12.1 Embedding of a non-reversible function

(a) Original function				(b) Degree of freedom						(c) One possible embedding					
x_0	x_1	y_0	y_1	x_0	x_1	a_0	y_0	y_1	g_0	x_0	x_1	a_0	y_0	y_1	g_0
0	0	0	0	0	0	0	0	0	*	0	0	0	0	0	0
0	1	0	1	0	0	1	.	.	.	0	0	1	1	1	0
1	0	1	0	0	1	0	0	1	*	0	1	0	0	1	0
1	1	0	1	0	1	1	.	.	.	0	1	1	1	1	1
				1	0	0	1	0	*	1	0	0	1	0	1
				1	0	1	.	.	.	1	0	1	0	0	1
				1	1	0	0	1	*	1	1	0	0	1	1
				1	1	1	.	.	.	1	1	1	1	0	0

In addition to a unique mapping from inputs to outputs, reversibility requires that the number of inputs and outputs have to be equal. Therefore, if n is larger than $m + k$, $n - m - k$ further garbage outputs are added and marked with *. In the opposite case, $m + k - n$ additional inputs (the so-called *ancillary inputs*) have to be added to the function. Each additional input doubles the number of rows in the truth table. If all ancillary inputs are assigned 0, the reversible function evaluates to the originally specified output. For all other assignments to the ancillary inputs, again arbitrary output values can be applied—even for the primary outputs. However, also here dependencies have to be considered. In fact, while the output pattern indeed is *don't care* in these cases, after all, each pattern is supposed to be applied only once in order to ensure reversibility. In the following, this is represented by a dot ($\cdot$).

Example 12.1 (Continued) *One ancillary input is needed to ensure that the number of inputs is equal to the number of outputs. This yields a truth table of the embedded function as shown in Table 12.1b (the original function is highlighted with bold entries). If the additional input a_0 is set to 0, the intended function can be obtained from the outputs y_0 and y_1. The values of the additional garbage output g_0 as well as for the remaining input assignments (i.e., for $a_0 \neq 0$) can be arbitrarily chosen (represented by * and $\cdot$, respectively) as long as the dependencies discussed above are considered.*

Finally, the embedding process is completed by assigning precise values to all entries represented by * and $\cdot$ while considering the discussed dependencies.

Example 12.1 (Continued) *Assigning * and $\cdot$ with precise values may yield the reversible function shown in Table 12.1c.*

The process reviewed above seems simple if small functions such as the half adder are considered. However, with increasing function size, embedding gets costly. In fact, it has been proven in [155] that embedding is co$\mathcal{NP}$-complete. Two main challenges exist: First, determining the most frequent output pattern and, hence, the number $k = \lceil \log_2 \mu(p_1) \rceil$ of additionally required garbage outputs requires a consideration of all possible output patterns in the worst case—an exponential complexity. Second, assigning all *- and $\cdot$-entries with precise values (while respecting the dependencies) is non-trivial.

In the past, embedding has mainly been conducted by means of truth tables [97]—obviously not feasible for large functions. Recently, researchers started investigations towards embedding of larger functions using BDDs. This led to improvements, e.g., in the determination of the minimum number of actually required garbage outputs [155, 174] as well as more elaborated embedding methods [155]. However, while exact embedding still remained intractable for large functions, the work in [155] also solved this problem heuristically, i.e., an embedding where the number of additional outputs is not necessarily the minimum. For these embeddings, the number of outputs is approximated by $n + m$, but it remains uncertain how far this if away from the actual minimum.

12.1.2 DD-Based Embedding

This section presents a DD-based method that allows for an exact embedding for larger functions. In contrast to previous approaches, which rely on truth tables (or representations of them like BDDs), this approach is based on function matrices. The provided methodology is divided into two steps: First, the minimum number of additionally required garbage outputs is determined. Then, the precise values for the respectively resulting *- and ·-entries are determined. Since function matrices can be represented compactly using DDs (cf. Chap. 11), this results in an efficient embedding process.

Determining the Number of Garbage Outputs

As discussed in Sect. 12.1.1, the number $\mu(p_1)$ of occurrences of the most frequent output pattern p_1 has to be determined first. If the considered function f is represented as a function matrix, $\mu(p_1)$ is equal to the maximum number of 1-entries in a row of this function matrix. Since all entries in the function matrix are either 1 or 0, counting the 1-entries is equivalent to calculating the row sum.

Example 12.2 *Consider the function matrix of the non-reversible function as shown in Fig. 12.1a. The row sum for output pattern* $p_1 = 01$ *is 2, since two input combinations (01 and 11) are mapped to this output pattern. The other row sums are either 0 or 1. Hence, the number of occurrences of the most frequent output pattern is* $\mu(p_1) = 2$.

Forming the row sums of a function matrix M turns out to be very simple: M is a square matrix composed of Boolean values, which contains exactly one 1-entry in each column. The product of M with its transposed[1] matrix M^T yields a diagonal matrix (i.e., a matrix where all entries off the main diagonal are zero) $D = M \cdot M^T$ with the row sums of M in its main diagonal.

Inputs

Outputs ($x_0 \backslash x_1$)	00	01	10	11
00	**1**	0	0	0
01	0	**1**	0	**1**
10	0	0	**1**	0
11	0	0	0	0

(a) M

×

Outputs

Inputs ($x_0 \backslash x_1$)	00	01	10	11
00	**1**	0	0	0
01	0	**1**	0	0
10	0	0	**1**	0
11	0	**1**	0	0

(b) M^T

=

$x_0 \backslash x_1$	00	01	10	11
00	**1**	0	0	0
01	0	**2**	0	0
10	0	0	**1**	0
11	0	0	0	**0**

(c) $D = M \cdot M^T$

Fig. 12.1 Determine the minimum number of additional outputs

[1] The transposed of a matrix is obtained by reflecting all elements along the main diagonal.

Example 12.3 *Figure 12.1a shows the function matrix M of a non-reversible function. Furthermore, Fig. 12.1b and c shows its transposed matrix M^T as well as the diagonal matrix $D = M \cdot M^T$, respectively. The product $M \cdot M^T$ contains the row sums of M, i.e., the number of occurrences of the output patterns, in its diagonal. Since no input maps to output 11 the fourth entry in the main diagonal is 0. The second entry of the diagonal is 2, since input pattern 01 as well as input pattern 10 maps to output 01. Since this is the largest entry, $p_1 = 01$ and $\mu(p_1) = 2$.*

This DD-based approach is fundamentally different from previous ones, because the matrix multiplication allows considering all output patterns concurrently. The conducted evaluation shows that it determines $\mu(p_1)$ efficiently, whereas previous approaches require substantial run-time. More precisely, an improvement of several orders of magnitude in terms of run-time is observed.

Assigning Precise Values

After determining $\mu(p_1)$, it is simple to correspondingly extend the considered function matrix M by the respective number of $k = \lceil \log_2 \mu(p_1) \rceil$ additional inputs and outputs: Each entry of M is simply replaced with a new $2^k \times 2^k$ matrix.[2] In order to keep the original function, the corresponding 1-entries and 0-entires of M are treated differently. More precisely, the 1-entries are replaced by a matrix B_1, whereas the 0-entries are replaced by a matrix B_0. The first column of the $2^k \times 2^k$ matrix B_1 contains *-entries only (representing that the desired output is determined when setting the additionally added ancillary inputs to zero). In contrast, the first column of the $2^k \times 2^k$ matrix B_0 contains 0-entries only (representing that the input pattern must not be mapped to one of these output patterns). All remaining columns of B_1 and B_0 are filled with ·-entries. After replacing the 1-entries and 0-entries with the corresponding matrices, it remains open how to assign the respective *- and ·-entries with actual values (while, at the same time, respecting the dependencies). How to do this is covered in this section. To this end, the remaining problem is illustrated first.

Example 12.4 *In the running example, the non-reversible function is extended by one additional input/output (since $k = \lceil \log_2 2 \rceil = 1$). Therefore, the 1-entries and 0-entries of M (cf. Fig. 12.1a) are replaced by $2^1 \times 2^1$ matrices B_1 and B_0, respectively, as shown in Fig. 12.2a (B_1-matrices are highlighted by dashed squares; all remaining 2×2-matrices are B_0-matrices). This ensures that, e.g., the desired input pattern $x_0 x_1 = 11$ still maps to the desired output pattern $x_0 x_1 = 01$ while, at the same time, the flexibility of whether the extended input pattern $x_0 x_1 a_0 = 110$ has to map to the extended output pattern $x_0 x_1 a_0 = 010$ or to the extended output pattern $x_0 x_1 a_0 = 011$ is still obtained. Now, the problem remains what output pattern shall be chosen so that, eventually, a unique input/output mapping is guaranteed.*

---

[2] If $m + k > n$, only $k + m - n$ inputs and outputs have to be added.

(a) Extended function matrix

Inputs

$\frac{a_0}{x_1 \ x_0}$	000	001	010	011	100	101	110	111
000	*	·	0	·	0	·	0	·
001	*	·	0	·	0	·	0	·
010	0	·	*	·	0	·	*	·
011	0	·	*	·	0	·	*	·
100	0	·	0	·	*	·	0	·
101	0	·	0	·	*	·	0	·
110	0	·	0	·	0	·	0	·
111	0	·	0	·	0	·	0	·

(b) Block diagonal matrix

Inputs

$\frac{a_0}{x_1 \ x_0}$	000	001	010	011	100	101	110	111
000	*	·	0	0	·	0	·	·
001	*	·	0	0	·	0	·	·
010	0	·	*	*	·	0	·	·
011	0	·	*	*	·	0	·	·
100	0	·	0	0	·	*	·	·
101	0	·	0	0	·	*	·	·
110	0	·	0	0	·	0	·	·
111	0	·	0	0	·	0	·	·

(c) Resulting embedding

Inputs

$\frac{a_0}{x_1 \ x_0}$	000	001	010	011	100	101	110	111
000	1	0	0	0	0	0	0	0
001	0	1	0	0	0	0	0	0
010	0	0	1	0	0	0	0	0
011	0	0	0	0	0	0	1	0
100	0	0	0	0	1	0	0	0
101	0	0	0	0	0	1	0	0
110	0	0	0	0	0	0	0	1
111	0	0	0	1	0	0	0	0

Fig. 12.2 Embedding of a non-reversible function using function matrices

Hence, the challenge is how to assign precise values to the *- and ··-entries such that a reversible function results, i.e., such that each row and each column of the *function matrix* contains a single 1-entry. A primitive solution would be to consider each row/column after each other. Again, this would tackle the exponential complexity in an enumerative and, thus, inefficient fashion. Hence, an alternative that focusing on the *-entries first (their assignment requires more dependencies to be considered) is considered. These dependencies are resolved by transforming the matrix into a block diagonal matrix with blocks of dimension $2^k \times 2^k$ along the main diagonal. This is achieved by swapping columns and results in a matrix where inherently all zero entries are located outside of the $2^k \times 2^k$ blocks along the main diagonal.

Example 12.5 *Consider again the extended matrix for the non-reversible function shown in Fig. 12.2a. By swapping columns 110 and 111 followed by swapping columns 111 and 011, a matrix as shown in Fig. 12.2b results, where all *-entries are located inside the 2×2 blocks (matrices) along the diagonal (highlighted by solid squares); all 0-entires are outside of these blocks.*

Recall that all *- and ··-entries have to be assigned in a reversible fashion (i.e., respecting the dependencies). Due to the last step, all these entries are now arranged along the main diagonal. Hence, the desired assignment can be easily obtained by employing the identity function, i.e., setting all entries along the diagonal to 1. This clearly yields a reversible function (in fact, the identity function is the simplest reversible function) and implicitly satisfies all dependencies. At the same time, the originally desired function is easily determined by reverting the column swaps conducted before (possible since swapping of columns is a reversible operation). This eventually yields a complete matrix for the desired function, which fully satisfies all dependencies and, hence, constitutes a full embedding.

Example 12.5 (Continued) *Employing the identity function to the matrix in Fig. 12.2b yields an assignment to all *- and ··-entries which satisfies all dependencies and, hence, is reversible. By swapping columns 011 and 111 followed*

by swapping columns 110 and 111, the matrix as shown in Fig. 12.2c results which, eventually, describes the original function in a reversible fashion.

Implementation

Thus far, this section has presented a complementary approach for embedding non-reversible functions into reversible ones by utilizing function matrices. This does not necessarily allow conducting the embedding process more efficiently since function matrices suffer from an exponential complexity. However, as discussed in Chap. 11, function matrices can be often represented very compactly using DDs—allowing to tackle the underlying complexity.

In order to conduct the embedding process as presented above, the following operations have to be conducted on a DD-based representing of the function matrix: matrix multiplication, transposition, determining the largest value in the diagonal, extending the function matrix, and block diagonalization. As already discussed in Sect. 5.2.2 multiplication can be realized directly (and, thus, efficiently) on the DD. Furthermore, also all other operations needed for the embedding approach are efficiently implemented on top of these DDs as follows:

- The *transposition* of a matrix M represented by a DD-node is realized as sketched in Fig. 12.3a. Here, the sub-matrices M_{01} and M_{10} are swapped and, afterwards, these swaps are recursively applied for all four sub-matrices.
- The *determination of the largest value* $\mu(p_1)$ *along the diagonal* of a matrix M is realized by a post-order traversal of the DD-nodes as sketched in Fig. 12.3b. Here, the maximum values μ_i of all sub-matrices are recursively determined and, afterwards, their maximum yields μ. The maximum value of a terminal node is its value.[3] Since the sub-matrices M_{01} and M_{10} are *zero matrices* in a diagonal matrix, only sub-matrices M_{00} and M_{11} have to be considered.
- *Extending the function matrix* is easy for DDs if matrices B_1 and B_0 are modeled as DDs as well. Then, all edges pointing to a 1- or 0-entry are replaced by edges pointing to B_1 and B_0, respectively. Without the loss of generality, all $\cdot$-entries are represented as 0, since they are not considered during block diagonalization. Hence, B_0 is modeled as a *zero matrix*. Of course, 0-entries must not be replaced by B_0, because a 0-terminal already represents a *zero matrix*. Furthermore, top most *-entry of B_1 is chosen to be set to 1, since the vertical position is changed after block diagonalization anyway. Thus, B_1 is represented by a DD, which contains exactly one node for each garbage output. This eventually yields the extension as sketched in Fig. 12.3c.[4]

---

[3] As discussed above, the DDs are implemented with a single terminal with value 1. Values other than 1 are represented by weighs attached to the edges. These weights serve as scalars the sub-matrices are multiplied with.

[4] Note that this process is a Kronecker multiplication of the original matrix with the corresponding B_1- and B_0- matrices.

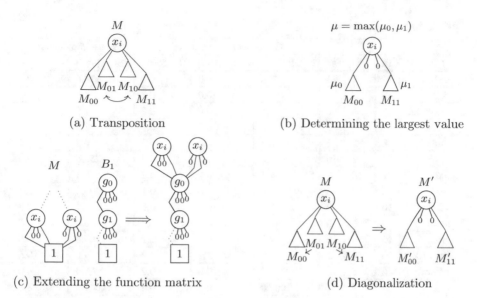

Fig. 12.3 Matrix operations on DDs

- *Block diagonalization* of DDs is accomplished by moving 1-entries from sub-matrices M_{01} and M_{10} to sub-matrices M_{00} and M_{11}, respectively. This is sketched in Fig. 12.3d. This movement can be realized by swapping columns with Hamming distance of 1, which can be implemented as matrix multiplication as shown in [154]. Afterwards, the algorithm is recursively applied to the sub-matrices M'_{00} and M'_{11} until all 1-entries are located inside $2^k \times 2^k$ blocks along the main diagonal.

12.1.3 Evaluation

The method as well as the implementation described above eventually yields an alternative embedding process, which is significantly more efficient than the state of the art. To confirm this, an evaluation has been conducted that compares the presented DD-based approach to previous solutions. To this end, the method described above has been implemented in C++ on top of the DD-package described in Chap. 7 and applied the same benchmarks to it as used in the previous work: the ESOP-based and BDD-based approach presented [155]. The conducted evaluation has been carried out on a 3.20 GHz Intel i5 processor with 8 GB of main memory running Linux 4.2—a machine, which is similar to what has previously been used in [155]. That is, the obtained run-times are comparable. Table 12.2 summarizes all results. The timeout was set to 5000 s.

Table 12.2 Evaluation

Name	PI	PO	k[a]	Garbage outputs determ.			Embedding determ.	
				ESOP [155]	BDD [155]	DD-based	ESOP [155]	DD-based
co14	14	1	14	72.8	0.0	0.2	118.6	0.2
dc2	8	7	6	0.1	0.0	0.2	0.1	0.2
example2	10	6	8	1.4	0.0	0.2	1.5	0.3
inc	7	9	5	0.0	0.0	0.2	0.0	0.2
mlp4	8	8	5	0.2	0.0	0.2	0.2	0.2
ryy6	16	1	16	157.7	0.0	0.2	258.8	0.3
5xp1	7	10	0	0.1	0.0	0.2	0.1	0.2
t481	16	1	16	1717.8	0.0	0.2	2308.4	2.3
x2	10	7	9	0.1	0.0	0.2	0.1	0.2
add6	12	7	6	46.1	0.1	0.2	50.0	0.6
cmb	16	4	16	7.1	0.0	0.2	104.8	0.6
ex1010	10	10	5	2.9	1.1	0.2	9.0	0.6
pcler8	16	5	16	28.1	0.0	0.2	68.3	4.0
tial	14	8	11	1007.0	0.2	0.3	1047.3	8.3
alu4	14	8	11	1270.3	0.1	0.3	1306.4	7.1
apla	10	12	10	0.0	0.0	0.2	0.5	0.5
f51m	14	8	11	556.5	0.2	0.3	595.7	6.9
cu	14	11	14	0.0	0.0	0.2	0.8	1.8
in0	15	11	14	1.5	0.1	0.2	26.5	57.1
0410184	14	14	0	1227.8	7.4	0.4	1229.2	0.4
apex4	9	19	7	1.0	25.0	0.3	39.8	0.6
misex3	14	14	14	160.7	17.5	0.3	929.7	29.7
misex3c	14	14	7	327.2	2.8	0.2	472.0	28.9
cm163a	16	13	12	625.4	0.0	0.2	633.5	6.1
bw	5	28	4	0.0	0.0	0.3	0.1	0.3
parity	16	1	15	TO	0.9	0.4	TO	1.4
cm150a	21	1	21	TO	0.1	2.7	TO	2.7
mux	21	1	21	TO	0.1	2.5	TO	2.6
cordic	23	2	23	TO	0.1	0.3	TO	3783.8
frg1	28	3	27	TO	0.0	0.3	TO	1522.3
pdc	16	40	15	31.1	TO	0.8	TO	1937.1
spla	16	46	15	32.7	TO	0.9	TO	738.7
ex5p	8	63	5	0.4	TO	1.2	TO	2.1
apex2	39	3	39	TO	5.1	1.1	TO	TO
e64	65	65	64	0.1	TO	1.5	TO	TO
seq	41	35	40	TO	TO	1.1	TO	TO
cps	24	109	23	TO	TO	4.5	TO	TO

PI primary inputs, *PO* primary outputs, $k = \lceil \log_2 \mu(p_1) \rceil$ Min. required garbage, *ESOP* Exact cube-based approach [155]; uses ESOP to det. k and an exact embedding, *BDD* BDD-based approach [155] to det. k, *DD-based* Approach presented in Sect. 12.1.2. The timeout was set to 5000 s
[a]Note that the total number of qubits is $\max(PI, PO + k)$

As discussed above, the state-of-the-art solutions often cannot tackle the exponential complexity of the embedding problem. In many cases, they already fail in the first step of the embedding process, i.e., when determining the minimum number of garbage outputs. This is reviewed in the fifth and sixth column of Table 12.2, which provide the run-time for the ESOP- and the BDD-based approach, respectively. In contrast, the presented DD-based solution can efficiently deal with this complexity—as shown in the seventh column of Table 12.2. While previous work often requires hundreds of CPU seconds or even times out in many cases, the solution presented in this chapter is capable of determining the number of garbage outputs in a few seconds or often in a fraction of a second only. Moreover, this allows obtaining exact values for the benchmarks *seq* and *cps*; thus far, only a heuristic number of required garbage outputs were known for these cases (cf. [174]).

Besides that, major improvements can also be observed for the second step of the embedding process (the assignment of precise values). The last two columns of Table 12.2 provide the values for the determination of an exact embedding (including the determination of the number of garbage outputs). Again, the run-time of the current state-of-the-art solution (i.e., the ESOP-based approach as proposed in [155]) as well as the run-time of the approach presented in Sect. 12.1.2 is provided.[5] Obviously, when already the determination of the number of garbage outputs failed, also no complete embedding can be created. Hence, no results could have been generated for all benchmarks that already timed out in the first step. Besides that, the state-of-the-art solution significantly suffers from large run-times in the second step as well (e.g., *pdc*, *spla*, *ex5p*, and *e64* timeout). Even though the presented DD-based method requires some time to tackle the exponential complexity—in some cases even timeouts are reported—significant improvements compared to the state of the art can be observed. In some cases, an embedding is obtained up to four orders of magnitude faster than with the state of the art.

12.2 Conducting Synthesis

The embedding process described above yields a reversible function f. Then, the synthesis task determines a reversible circuit $G = g_0 g_1 g_2 \ldots g_{|G|-1}$ (cf. Sect. 2.3) which realizes this function f. Following the general idea of functional synthesis, this can be conducted by applying reversible gates g_i to f so that, eventually, the identity function I results. Assume that applying a cascade of reversible gates $G^{-1} = g_{|G|-1} \ldots g_1 g_0$ transforms f into I. Then, due to the reversibility ($f \circ f^{-1} = I$), the inverse cascade $(G^{-1})^{-1} = G = g_0 g_1 \ldots g_{|G|-1} h$ realizes f (since the Toffoli gates g_i in self-inverse).

---

[5]Note that the BDD-based approach presented in [155] is not capable of conducting the second embedding step in an exact fashion and, hence, is not considered here.

Like the embedding step, also synthesis suffers from an exponential complexity. However, relying on permutation matrices that describe the embedded function to be realized (which again are represented by means of DDs) allows conducting this process in an efficient fashion. An initial approach following this idea has been proposed in [154]. This section (based on [191]) shows that this approach can be significantly improved by exploiting further redundancies in the paths and the nodes of the DDs—leading to a cost reduction of the synthesized circuit by several orders of magnitudes in many cases.

In the following, Sect. 12.2.1 reviews the general idea of the DD-based synthesis approach described in [154], while Sect. 12.2.2 describes improvements to this scheme.

12.2.1 DD-Based Synthesis

Following the general idea of the originally proposed DD-based functional synthesis approach, the goal is to find a sequence of gates that transform a DD (compactly representing a permutation matrix M describing a reversible function) to the identity I. Since this identity matrix I only represents mappings from 0 to 0 and from 1 to 1, each node in the DD has to be transformed such that its second edge and third edge point to a 0-stub (as illustrated in Fig. 12.4 for a node labeled x_i). Hence, for a given DD representing M, the task remains how to apply reversible gates such that this structure results eventually.

This task is addressed by successively transforming the DD towards the identity. To this end, the nodes of the DD are considered in a breadth-first traversal from the top to the bottom. In each step, the currently considered node (representing the partition according to variable x_i) is transformed into the desired structure. This is accomplished by applying Toffoli gates which move all 1-paths (i.e., all paths from the currently considered node to the 1-terminal) of the second and third edge to the first and fourth edge—eventually leading to nodes as illustrated in Fig. 12.4.

The main principle is to use Toffoli gates to swap DD-paths. More precisely, applying a gate $TOF(C, x_i)$ to a given DD inverts the input of the mapping of variable x_i for all paths represented by C (cf. Definition 2.6 on Page 19). This way, Toffoli gates may be used to swap, e.g., a mapping from 0 to 1 to a mapping from 1 to 1 and, by this, moving 1-paths from the third edge to the fourth edge—bringing it closer to the identity structure. An example illustrates the idea.

Fig. 12.4 Identity structure

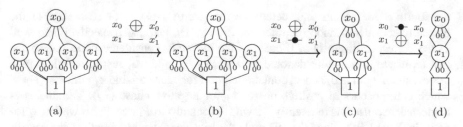

Fig. 12.5 Effects of applying Toffoli gates to DDs

Example 12.6 *The gate $TOF(\emptyset, x_0)$ inverts the value of the input of variable x_0 for all paths and, therefore, simply exchanges the first (third) and the second (fourth) edge of the root node of the DD shown in Fig. 12.5a. The resulting DD is depicted in Fig. 12.5b. The gate $TOF(\{x_1^+\}, x_0)$ inverts the input of variable x_0, but only for paths where variable x_1 maps from 1 to anything. This already yields the identity structure for the root node of the DD as shown in Fig. 12.5c.*

In addition, it has to be ensured that applying Toffoli gates does not affect previously traversed nodes. To this end, further control lines are added, which describe the path to the currently considered node to each Toffoli gate that is applied. More precisely, if the path to the currently considered node traverses the first edge of another node (representing a mapping from 0 to 0), a negative control line is added for the corresponding variable. Analogously, a positive control line is added for the corresponding variable if the path traverses the fourth edge of that node.[6] If there exist k such paths to the currently considered node, each Toffoli gate is replicated for each of those k paths in order to eventually transform the currently considered node to the identity structure.

Example 12.6 (Continued) *Consider the DD shown in Fig. 12.5c and assume that the right DD-node with label x_1 is currently processed. Each gate applied for processing this node has to include a positive control line x_0. As a result, the gate $TOF(\{x_0^+\}, x_1)$ only swaps paths with $x_0 = 1$, i.e., paths which run through the fourth edge of the top node (representing a mapping from 1 to 1)—resulting in the identity DD as shown in Fig. 12.5d.*

Following the main principle outlined above and assuming that, without loss of generality, the currently considered node is labeled with variable x_i ($0 \leq i < n$) as well as the fact that all previously traversed nodes (i.e., all nodes labeled with variable x_l, where $0 \leq l < i$) already establish the identity structure, the currently considered node can be transformed to the identity structure as follows:

Apply Toffoli gates such that all 1-paths are moved from the second to the first edge while, at the same time, all 1-paths are moved from the third edge to

---

[6]A path to the currently considered node can only traverse the first or the fourth edge of other nodes, because they already establish the identity structure.

the fourth edge. To this end, determine the sets of 1-paths for each edge of the currently considered node (denoted by P_1, P_2, P_3, and P_4, respectively) as well as the corresponding sets of 0-paths of the currently considered node (i.e., paths that terminate in a 0-stub; denoted by $\overline{P}_1$, $\overline{P}_2$, $\overline{P}_3$, and $\overline{P}_4$, respectively). A path represents an input and, hence, contains a literal for each variable x_j with $i < j < n$ which either occurs in positive phase (x_j) or negative phase ($\overline{x}_j$). Variable x_i is neglected, because it is inherently known ($\overline{x}_i$ for paths in P_1 and P_3, as well as x_i for paths in P_2 and P_4). Since the DD-node describes a reversible function, the sets P_1 and P_3 are disjoint (both represent a mapping with input $x_i = 0$), i.e., $P_1 \cap P_3 = \emptyset$. Moreover, the set $\overline{P}_1$ of 0-paths through the first edge is equal to the set P_3 of 1-paths through the third edge. Finally, due to reversibility, the cardinalities of the sets P_2 and $\overline{P}_1 = P_3$ are equal.

Example 12.7 *Consider again the running example of Sect. 12.1, i.e., the non-reversible function shown in Table 12.1a on Page 107 as well as its embedding into a reversible function as shown in Table 12.1c. The corresponding DD representing the permutation matrix of this embedding is shown in Fig. 12.6a. Consider the top node of this DD. The sets of 1-paths are $P_1 = \{\overline{x}_1\overline{x}_2, x_1\overline{x}_2\}$, $P_2 = \{\overline{x}_1x_2, x_1\overline{x}_2\}$, $P_3 = \{\overline{x}_1x_2, x_1x_2\}$, and $P_4 = \{\overline{x}_1\overline{x}_2, x_1x_2\}$. Note that $P_1 \cap P_3 = P_2 \cap P_4 = \emptyset$ and that $\overline{P}_1 = \{\overline{x}_1x_2, x_1x_2\} = P_3$ are the 0-paths through the first edge.*

Because of the relation between 1-paths and 0-paths discussed above, each 1-path of the second edge can be swapped with a 0-path of the first edge. To keep the number of required Toffoli gates as small as possible, a 1-path $p \in P_2$ is swapped with its most similar 0-path $p' \in \overline{P}_1$. In fact, if there exist corresponding paths p and p' which are identical ($p = p'$), only a Toffoli gate with target line x_i and a set of control lines that represent p is required.[7] If $p \neq p'$, p has to be adjusted to match p' before the paths can be swapped as described above. To this end, a Toffoli

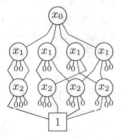

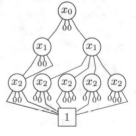

(a) Before processing the root node   (b) After processing the top node

Fig. 12.6 DD-based synthesis of the embedded function shown in Table 12.1c

---

[7]As discussed above, the set of control lines is additionally enriched with literals that specify the path to the currently considered node.

Fig. 12.7 Circuit obtained by the established design flow

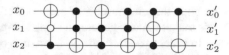

gate with target line x_j is added for each variable x_j that occurs in p and p' in different phases. Note that all these Toffoli gates contain a positive control line x_i^+, since only the paths in the second and fourth edge shall be changed. Swapping all 1-paths of the second edge with the 0-paths of the first edge inherently swaps the 1-paths of the third edge with the 0-paths of the fourth edge and, hence, transforms the currently considered node to the identity structure.

Example 12.7 (Continued) *Consider again the top node of the DD depicted in Fig. 12.6a. As can be seen, there exists a 1-path $p \in P_2$ which is identical to a 0-path $p' \in \overline{P}_1$, namely $p = \overline{x}_1 x_2 = p'$. To swap the 1-path with the 0-path, a gate $TOF(\{x_1^-, x_2^+\}, x_0)$ is applied. The remaining 1-path $q = x_1 \overline{x}_2$ of P_2 has to be adjusted to match the 0-path $q' = x_1 x_2$. This requires a gate $TOF(\{x_0^+, x_1^+\}, x_2)$. Since afterwards $q = q'$, the paths are swapped using a gate $TOF(\{x_1^+, x_2^+\}, x_0)$— resulting in the DD shown in Fig. 12.6b where the currently considered node assumes the identity structure. Continuing these steps for all remaining nodes eventually yields the circuit shown in Fig. 12.7a realizing the function represented by the DD in Fig. 12.6a (and, hence, the function described in Table 12.1a on Page 107 in reversible logic).*

12.2.2 Exploiting Redundancies in Paths and Nodes

In the DD-based synthesis scheme originally proposed in [154] and reviewed above, the overall number of paths to the nodes labeled with a certain variable x_i grows exponentially with the number of variables x_j with higher priority (i.e., $x_j \succ x_i$). This leads to a substantial number of gates with (partially) redundant sets of control lines—posing a significant drawback since

- a significant number of gates is applied to transform each single node of the considered DD to the identity structure and
- the applied gates usually include rather large sets of control lines.

More precisely, the number of gates is heavily influenced by the number of paths to the currently considered node, because each gate that is required to transform the currently considered node to the identity has to be replicated for each path. Furthermore, the number of control lines of these Toffoli gates depends on the number of literals of the path from the root node to the currently considered node (because these literals are added to the Toffoli gates in form of control lines to ensure that no other node is affected). Since the overall costs of a reversible circuit depend on both, the total number of gates as well as the respective number of control lines, this makes circuits generated using DD-based synthesis rather expensive.

In order to address the problem, two optimization techniques to reduce the costs of the circuits generated by DD-based synthesis are considered, namely

1. a straightforward solution which performs logic minimization on the paths to the currently considered node to reduce the number of paths as well as the number of their literals and
2. a more elaborate approach which considers nodes that require the same sequence of Toffoli gates in order to get transformed to the identity structure jointly.

The straightforward solution utilizes logic minimization techniques to reduce the overall number of paths to the currently considered node as well as to reduce the overall number of literals in the paths. To this end, each path to the currently considered node is described as a product (conjunction) of its literals. Then, the exclusive sum (exclusive disjunction) of all these products is formed. The sum has to be exclusive, because applying a Toffoli gate an even number of times does not have any effect (since a Toffoli gate is self-inverse). The resulting *Exclusive-Or Sum-of-Products* (ESoP) can be minimized using techniques such as proposed in [108]. Such a minimization reduces the overall number of products (and, hence, the number of paths and gate replications) as well as the number of literals in these products (and, hence, the number of control lines that have to be added to each gate).

Example 12.8 *Consider the DD depicted in Fig. 12.8 and assume that the node highlighted in blue is currently considered. This node can be transformed to the desired identity structure by applying a Toffoli gate with target line x_2. Since there exist two paths to this node, namely $\overline{x}_0 x_1$ and $x_0 \overline{x}_1$, two gates $TOF(\{x_0^- x_1^+\}, x_2)$ and $TOF(\{x_0^+ x_1^-\}, x_2)$ are applied to eventually transform this node to the desired identity structure (following the original DD-based synthesis algorithm reviewed in Sect. 12.2.1). The resulting circuit is shown in Fig. 12.9a.*

However, forming the ESoP of the two paths (i.e., $x_1 \overline{x}_2 \oplus \overline{x}_1 x_2$) allows exploiting redundancies in the paths to the currently considered node. Minimizing this ESoP yields $x_0 \overline{x}_1 \oplus \overline{x}_0 x_1 = x_0 \oplus x_1$. Consequently, also the two paths x_0 and x_1 might be used to describe all paths to the currently considered node. The resulting gates are shown in Fig. 12.9b. Although the number of paths (and therefore the number of

Fig. 12.8 Paths to the
currently considered node

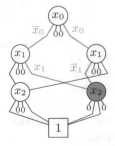

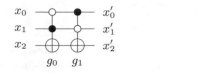

(a) Without optimization		(b) ESoP-minimized

Fig. 12.9 Gates required to transform the currently considered node from Fig. 12.8

Fig. 12.10 DD-nodes with
equal sets P_1 and P_2

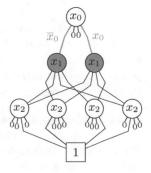

*gates) did not change, the costs of the circuits (e.g., T-count) reduce since the gates
have fewer control lines.*

The more elaborated optimization approach aims to further reduce the costs
of the circuits considering more than one node simultaneously. The general idea
is based on the key observation that different DD-nodes might require the same
sequence of Toffoli gates in order to get transformed to the identity. As described
in Sect. 12.2.1, this sequence of Toffoli gates depends on the set of 1-paths through
the first edge P_1 and the set of 1-paths through the second edge P_2 only. These two
sets uniquely determine the other sets of 1-paths ($P_3 = \overline{P_1}$, $P_4 = \overline{P_2}$) as well as
the sets of 0-paths. Cases frequently occur where nodes in the DD have equal sets
of 1-paths P_1 and P_2, even though they are structurally different.

Example 12.9 *Consider the DD depicted in Fig. 12.10. The root node already
establishes the desired identity structure and the two nodes labeled x_2 (highlighted
in blue) are structurally different. However, their sets of 1-paths are equal, i.e., $P_1 =
\{\overline{x}_2\}$ and $P_2 = \{\overline{x}_2\}$ for both nodes. Consequently, both nodes can be transformed to
the identity structure with the same sequence of Toffoli gates. One possible sequence
is $TOF(\{x_1^+\}, x_2), TOF(\{x_2^+\}, x_1)$.*

*Without applying this scheme, the sequence of Toffoli gates has to be replicated
twice—once for each node (including control line x_0^- for the left node and control
line x_0^+ for the right node). This resulting in the circuit shown in Fig. 12.11a. The
gates g_0 and g_1 thereby transform the left node to the identity, whereas the gates g_2
and g_3 transform the right node to the identity.*

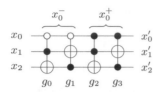

(a) Without joint consideration

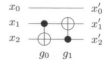

(b) With joint consideration

Fig. 12.11 Gates required to transform the nodes labeled x_1 in Fig. 12.10

Since DD-nodes that employ an equal characteristic regarding their sets of 1-paths P_1 and P_2 can be transformed to the identity structure with the same sequence of Toffoli gates, they are considered jointly for synthesis purposes and thus processed together. To this end, the ESoP of the paths to all nodes with equal sets P_1 and P_2 is formed and minimized as described in the straightforward approach.

Example 12.9 (Continued) *Since both nodes labeled x_2 have equal sets of 1-paths P_1 and P_2, they are considered jointly and processed together. The minimized ESoP of all paths to these nodes is $x_1 \oplus \overline{x}_1 = 1$, a sum consisting of a single product without any literals. Therefore no additional control lines are required (all nodes labeled x_2 are considered jointly). The resulting circuit that transforms all nodes labeled x_2 to the identity structure is shown in Fig. 12.11b. Compared to the gate sequence depicted in Fig. 12.11a, a reduction of the T-count from 28 to 0 is observed.*

12.2.3   Optimization Potential

This section briefly analyzes the potential of the optimization scheme described above, i.e., how many nodes might be considered together in the best case. To this end, it is assumed that the nodes of n variables are left to be processed, i.e., the currently considered nodes are sub-DDs with height n, and that the sequence of Toffoli gates that transforms these nodes to the identity structure is uniquely determined by the set of 1-paths P_1 and P_2.

First, it is determined how many different sequences of Toffoli gates exist. Assuming that the sequence of Toffoli gates is uniquely determined by P_1 and P_2 allows analyzing how many combinations of sets P_1 and P_2 exist: The DD of the currently considered node represents a $2^n \times 2^n$ permutation matrix (since all nodes above already employ the identity structure). Consequently, there must be exactly one 1-entry in each of the 2^n rows and in each of the 2^n columns. This means, that there must be exactly 2^{n-1} 1-entries in the upper half of the matrix, i.e., $|P_1| + |P_2| = 2^{n-1}$. These 2^{n-1} 1-entries (1-paths) are arbitrarily distributed in the 2^n columns (in the sets P_1 and P_2). Consequently, there exist

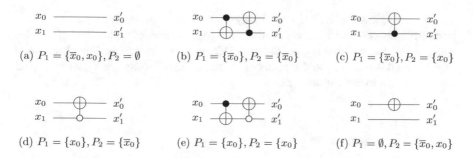

Fig. 12.12 Sequences of Toffoli gates for $n = 2$

Table 12.3 Potential of the considered optimization

n	No. sequences $\binom{2^n}{2^{n-1}}$	No. DDs $2^n!$
2	6	24
3	70	40,320
4	12,870	$2 \cdot 10^{13}$
5	$6 \cdot 10^8$	$2.6 \cdot 10^{35}$

$\binom{2^n}{2^{n-1}}$ possibilities in which rows the 1-entries are located, i.e., possible different pairs of sets (P_1, P_2). Assuming $n = 2$, there exist $\binom{2^2}{2^2-1} = 6$ different sequences that transform a currently considered node to the identity. These sequences as well as their corresponding sets P_1 and P_2 are depicted in Fig. 12.12.

As a second step, it is analyzed how many different sub-DDs with n variables exist. Recall that a DD composed of n variables represents a permutation matrix of dimension $2^n \times 2^n$, i.e., a matrix that represents a permutation of 2^n elements. Since 2^n elements can be permuted in $2^n!$ ways, there exist $2^n!$ structurally different DDs with n variables. Considering again that $n = 2$, there exist $2^2! = 4! = 24$ structurally different DDs.

Having an arithmetic expression for the number of sequences as well as for the number of DDs allows analyzing the potential of the considered optimization. The resulting numbers for several values of n are provided in Table 12.3. As one can easily see, there are many more different DDs than sequences, because $\binom{2^n}{2^{n-1}} \ll 2^n!$. Consider the case that the $n = 3$ variables of the DD are not yet processed. In the worst case, the DD has 40,320 nodes labeled with the currently considered variable. For each of these nodes, a sequence of Toffoli gates has to be determined. If the considered optimization is applied (i.e., if nodes with equal sets of 1-paths are jointly considered), the number of nodes that have to be processed drops to 70—reducing the computational effort by a factor of 576. Furthermore, the logic minimization used to reduce the paths to the currently considered nodes is applied to a larger set of paths, which makes it more likely to obtain a more compact ESoP.

Obviously, it is more likely that many nodes can be processed together if their currently considered variable is the label for a large number of nodes. Therefore, this optimization has a higher impact on large DDs than on small ones. The observation that large DDs tend to yield circuits with rather high T-count suggests to expect higher improvements for these cases.

12.2.4   Evaluation

This section evaluates the cost reduction of the synthesized circuit achieved by applying the considered optimizations to the DD-based synthesis algorithm. To this end, the DD-based synthesis approach originally proposed in [154] has been reimplemented (including some minor optimizations regarding performance) in C++ using the DD-package developed in Chap. 7 and the BDD-package CUDD [156]. This implementation represents the current state of the art and serves as baseline for the conducted evaluation. Based on that, the optimizations discussed in Sect. 12.2.2 have been implemented. In the following, only using logic minimization[8] (to reduce the number of paths) is denoted Scheme A, whereas additionally processing nodes with equal sets of 1-paths jointly is denoted Scheme B. As benchmarks served the reversible functions provided in RevLib [173] by means of reversible circuits. The evaluation has been conducted on a 4 GHz processor with 32 GB of memory running Linux 4.4.

Table 12.4 summarizes the conducted evaluation. The first two columns list the name of the benchmark and the number of circuit lines n. Then, for each synthesis scheme, the run-time as well as the $T\text{-}count$ is listed. Finally, the reduction of the costs for Scheme A with respect to the baseline and for Scheme B with respect to Scheme A are listed in the last two columns of Table 12.4.

The obtained results clearly show a significant improvement in terms of run-time. While the original approach requires a significant amount of run-time for some benchmarks (e.g., more than 1000 s for benchmarks $sym9$, $rd84$, and $cycle10$), the considered optimizations allowed to synthesize all benchmarks within a few seconds (Schemes A and B). Only one benchmark ($cordic$) required slightly more than a minute.

Furthermore, a substantial improvement in terms of costs of the synthesized circuit is observed for the benchmarks. Since the considered optimizations only focus on redundancies in paths and nodes of the DD in order to reduce the number of repetitions of gates sequences as well as to reduce the number of control lines, the observed improvements are purely attributed to them and not to lucky local choices. For all benchmarks that result in a circuit with a T-count of more than half a million using the original approach, substantial improvements of several orders of magnitude are achieved. Consider, for example, the benchmarks $plus127mod8192$

---

[8]The methods available at RevKit [149] are utilized for logic minimization.

Table 12.4 T-count improvements compared to the state of the art

| Benchmark | n | Baseline [154] | | Scheme A | | | Scheme B | | | Improvement | |
		t	T-count	t	T-count	t	T-count		Base/A	A/B
alu1	20	0.02	180	0.03	180	0.04	180		1.00	1.00
cmb	20	0.00	864	0.01	864	0.00	864		1.00	1.00
cycle17_3	20	0.00	30,105	0.01	30,105	0.01	27,852		1.00	1.08
ex1010	20	1.91	83,724	1.76	84,156	1.64	81,648		0.99	1.03
C7552	21	0.01	576	0.00	576	0.01	576		1.00	1.00
decod	21	0.00	576	0.00	576	0.01	576		1.00	1.00
dk17	21	0.01	1371	0.00	1371	0.01	1371		1.00	1.00
ham7	21	0.06	1,516,800	0.03	99,147	0.03	696		15.30	142.45
pcler8	21	0.00	300	0.00	300	0.00	300		1.00	1.00
alu4	22	6.66	116,109	1.89	122,946	1.95	118,038		0.94	1.04
apla	22	0.02	2652	0.02	2652	0.03	2652		1.00	1.00
cm150a	22	0.19	768	0.19	768	0.19	768		1.00	1.00
f51m	22	1.48	36,630	1.46	36,630	1.49	36,630		1.00	1.00
mux	22	0.19	768	0.19	768	0.18	768		1.00	1.00
tial	22	1.26	64,665	1.21	64,665	1.30	59,169		1.00	1.09
plus63mod4096	23	5.71	104,955,648	0.59	1,093,500	0.32	276		95.98	3961.96

(continued)

Table 12.4 (continued)

Benchmark	n	Baseline [154]		Scheme A		Scheme B		Improvement	
		t	T-count	t	T-count	t	T-count	Base/A	A/B
add8	25	7.92	452,175	3.90	151,512	4.00	151,512	2.98	1.00
cordic	25	76.91	448,080	72.43	448,080	73.42	448,080	1.00	1.00
cu	25	0.01	915	0.02	915	0.01	915	1.00	1.00
plus127mod8192	25	21.83	353,272,704	1.78	2,432,892	1.24	303	145.21	8029.35
plus63mod8192	25	21.91	353,272,704	1.85	2,432,892	1.23	303	145.21	8029.35
rd73	25	39.61	628,301,952	0.65	710,868	0.12	8 334	883.85	85.30
in0	26	0.44	15,057	0.46	15,057	0.44	15,057	1.00	1.00
sym9	27	1572.09	25,715,996,160	2.75	3,811,839	1.13	318,615	6746.35	11.96
apex4	28	2.25	58,641	2.12	58,641	2.22	58,005	1.00	1.01
cm151a	28	0.03	744	0.04	744	0.06	744	1.00	1.00
hwb5	28	88.62	1,056,981,792	2.41	1,737,033	1.91	11,556	608.50	150.31
misex3	28	2.22	92 991	2.24	92,991	2.29	92,931	1.00	1.00
misex3c	28	2.26	92,991	2.27	92,991	2.32	92,931	1.00	1.00
table3	28	2.22	59,112	2.16	59,112	2.15	59,244	1.00	1.00
cm163a	29	0.02	678	0.01	678	0.02	678	1.00	1.00
in2	29	0.15	19,296	0.16	19,344	0.15	19,296	1.00	1.00
frg1	31	2.50	15,219	2.62	15,219	3.06	15,219	1.00	1.00
mod5adder	32	446.53	6,142,116,000	1.72	1,047,696	1.30	12,453	5862.50	84.13
rd84	34	MO	–	1.25	1,019,451	0.92	16,794	–	60.70
cycle10	39	MO	–	6.24	6,803,379	5.71	1500	–	4535.59

n number of qubits t run-time in seconds, *Baseline* original approach as described in [154], *Scheme A* exploiting redundancies in paths, *Scheme B* additionally exploiting redundancies in nodes and *plus63mod8192*. Performing logic optimizations on the paths to the currently considered node (i.e., Scheme *A*) already results in a reduction of the T-count by a factor of 145.21. Additionally transforming nodes together that have equal sets of 1-paths (i.e., Scheme *B*), another improvement by a factor of 8029.35 and, hence, an overall improvement of six orders of magnitude is gained. On average, an improvement by a factor of 4.22 for Scheme *A* with respect to the original approach and an improvement by a factor of 5.35 of Scheme *B* with respect to Scheme *A* are observed. This results in an overall improvement by a factor of 22.57.

Chapter 13
One-Pass Design Flow

Abstract This chapter introduces a new design flow for Boolean components used in quantum computations termed *one-pass design*, which combines the embedding and the synthesis step while keeping the number of additionally required qubits at the minimum. This new design flow allows improving scalability and reducing costs of the resulting circuit since a much larger degree of freedom can be exploited for an optimized synthesis. Moreover, this new design flow is not tightly bound to a single synthesis approach, but can be applied to other functional synthesis approaches as well.

Besides that, the proposed one-pass design flow also allows utilizing coding techniques to reduce the number of additionally required qubits (for embedding). More precisely, it is shown that a single additional qubit is always sufficient when assuming a Huffman-like encoding and that no decoding is required afterwards (i.e., if subsequent components can handle encoded inputs). In cases where an additional decoder is required, this technique is still beneficial since the coded function is realized more efficiently in many cases.

Keywords One-pass design · Synthesis · Embedding · Coding · Qubits · Reversible circuits

While the improvements on embedding and DD-based synthesis discussed in the previous chapter already lead to a significant improvement in the design of reversible circuits with respect to efficiency and scalability, the currently established design flow still suffers from the need to conduct embedding and actual synthesis separately. More precisely, this disjoint consideration leads to the following main drawbacks:

- An embedding function is determined rather arbitrarily and without considering the following synthesis step. As an example, there are 192 possibilities on how to assign precise values for the ∗- and the ·-entries of the function considered in Example 12.1 on Page 108. But eventually the embedding process just picks a single solution without considering which embedding might be particularly suited for the synthesis steps to be conducted afterwards. This way, a huge degree of freedom is not exploited (since all 16 *don't cares* are already assigned prior to synthesis).

- Embedding a function almost always requires garbage outputs and, hence, constant inputs which, in turn, lead to an exponential increase in the truth table and/or function matrix description. Consequently, the actual synthesis is performed on a function representation which is usually much more complex than the representation of the originally intended target function. This poses a threat to the efficiency and scalability of the synthesis process.

This chapter (based on [193, 196]) addresses these drawbacks by introducing a new design flow for functional synthesis of reversible circuits that combines embedding and the synthesis step. This *one-pass design flow* allows exploiting a large degree of freedom since the *don't cares* in the embedding process (cf. Sect. 12.1) are assigned in a fashion that suits best to the considered synthesis approach— significantly improving scalability of functional synthesis and reducing the costs of the resulting circuits.

Moreover, this chapter (based on [188, 195, 197]) introduces the idea of enriching this new design flow with coding techniques. This allows reducing the number of variables considered during synthesis and—assuming that subsequent components are capable of handling encoded inputs—to reduce the number of additional qubits below what is currently considered to be the minimum. In fact, it is shown that a single ancillary qubit is sufficient to embed any non-reversible function in a reversible one by using encoded output patterns. Even in the case that decoding is required afterwards (increasing the number of required qubits to the minimum again) utilizing coding is beneficial since cheaper circuits result.

In the following, the one-pass design flow is presented. To this end, Sect. 13.1 describes this design flow using DD-based synthesis as an example. However, the presented one-pass design flow is not tightly bound to a specific synthesis scheme and, in fact, can be employed to many other functional synthesis approaches as well. Afterwards, Sect. 13.2 discusses the idea of utilizing coding techniques in order to reduce the number of required qubits and the costs of the synthesized circuits.

13.1   Combining Embedding and Synthesis

This section introduces the concept of a one-pass design flow. First, the importance of the embedding process is discussed, i.e., it is analyzed what problems arise when functional synthesis approaches are applied without a prior embedding process. Afterwards, two complementary solutions overcoming these problems and, hence, realizing one-pass design of reversible circuits are presented: The first one (introduced in Sect. 13.1.2) guarantees the minimum with respect to the number of required circuit lines, but addresses only the first drawback (exploiting the full degree of freedom). The second one (introduced in Sect. 13.1.3) also addresses the second drawback (keeping the function representation small), but may require a slightly larger (but still bounded) number of additional circuit lines. Both solutions can be incorporated to various functional synthesis schemes. In the following, DD-based synthesis (cf. Sect. 12.2.1) serves as a representative. The conducted evaluation (summarized later in Sect. 13.1.4) clearly shows that the presented *one-pass* design of reversible circuits addresses the drawbacks discussed above. In fact, for DD-based synthesis a significant reduction of the costs of the resulting circuits (up to several orders of magnitude) is observed. Besides that, a speedup of several orders of magnitude is observed—allowing to synthesize some of the frequently considered benchmarks with a minimum number of circuit lines for the first time.

13.1.1 Importance of the Embedding Process

This section analyzes what problems arise when functional synthesis approaches are applied without a prior embedding process. To this end, DD-based synthesis as discussed in Sect. 12.2 serves as representative to show why functional synthesis approaches fail when they are applied to non-reversible functions. These findings serve as basis for the *one-pass design flow* and allow adjusting functional synthesis approaches to follow this new design flow (again, DD-based synthesis serves as representative).

Recall DD-based synthesis approach reviewed in Sect. 12.2. The main idea is to swap the 1-paths of the second and third edge with the 0-paths of the first and fourth edge, respectively. This only works if $\left|\overline{P_1}\right| \geq P_2$ and $\left|\overline{P_4}\right| \geq P_3$ for the currently considered node, i.e., there must be at least as many 0-paths through the first edge as 1-path through the second edge. This is always the case if the function to be synthesized is reversible, because each output pattern occurs exactly once. However, in case the function to be synthesized is non-reversible (because the embedding step has been skipped), DD-nodes will occur for which the set of 0-paths through the first edge $\overline{P}_1$ contains fewer paths than the set of 1-paths through the second edge P_2, i.e., $\left|\overline{P}_1\right| < |P_2|$. Then, not all 1-paths of the second edge can be moved to the first edge, i.e., the identity structure cannot be established for the node. Consequently, the synthesis approach fails.[1]

Example 13.1 *Consider the top node of the rightmost DD shown in Fig. 13.2 representing the function matrix of the non-reversible function shown in Table 12.1a. The respective set of paths are $P_1 = \{\overline{x}_1, x_1\}$, $P_2 = \{x_1\}$, $P_3 = \emptyset$, and $P_4 = \{\overline{x}_1\}$. Since $\overline{P}_1 = \emptyset$ does not contain any 0-paths that can be swapped with the path of P_2, the synthesis approach cannot be applied as described in Sect. 12.2. Consequently, there is no chance to transform the currently considered node to the identity structure and, hence, the synthesis algorithm fails.*

---

[1]Note that similar problems arise when other synthesis approaches such as, e.g., [148] are applied without embedding.

The following sections describe how to overcome this problem—utilizing DD-based synthesis as representative. The general idea of the approach described in Sect. 13.1.2 is to add further 0-paths such that synthesis can be conducted (almost) as usual. To ensure that enough such paths are inserted, the minimal number of additionally required variables is determined (cf. Sect. 12.1). Since the number of required circuit lines is guaranteed to be the minimum here, this is denoted the *exact solution*. Even though the exact approach allows exploiting a significant degree of freedom, the drawback remains that the additionally required variables are added to the DD (thus, increasing its complexity). To overcome this issue, Sect. 13.1.3 presents an approach that does not add variables to the function f to be synthesized, but rather modifies f such that synthesis can continue without running into the problems discussed above. In order to restore the modifications on the function later, additional lines are utilized that buffer any changes to the function. This may require more circuit lines than the minimum number (although still bounded), which is why this is denoted the *heuristic solution*. Both (complementary) schemes conduct embedding *during* the synthesis rather than prior to synthesis—leading to the desired one-pass design of reversible circuits.

13.1.2   Exact Solution

To overcome the problem with the mismatching cardinalities of $\overline{P}_1$ and P_2 discussed above, the exact solution increases the overall number of 0-paths in the DD M without increasing the number of 1-paths. To this end, $k' = (m + k) - \max(n, m)$ garbage variables are added to the DD prior to synthesis (if k' is not negative; otherwise no further variables are added). These garbage variables build up a DD G as illustrated in Fig. 13.1, which has a single 1-path, namely $p = \overline{g}_0\overline{g}_1 \cdots \overline{g}_{k'-1}$. The garbage variables are added to M by replacing its terminal node with the root node of G (i.e., forming the Kronecker product $M \otimes G$). Consequently, the original function is obtained if all ancillary inputs are set to 0 (similarly as discussed in Sect. 12.1). For all other combinations of the ancillary inputs, the output is *don't care*. These *don't cares* are represented by 0-paths, since they do not have to be considered during synthesis.

Fig. 13.1 DD representing the garbage variables

Fig. 13.2 Extending the function matrix

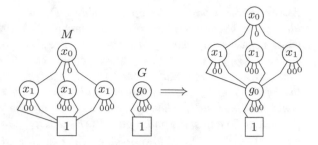

Example 13.2 *The left-hand side of Fig. 13.2 shows the DD M representing the function matrix of the non-reversible function provided in Table 12.1a on Page 107. Since $n = m = 2$ and $k = 1$, $k' = 2 + 1 - 2 = 1$ additional variable—named g_0—is added to the DD. The extended DD is shown on the right-hand side of Fig. 13.2.*

Adding variables to the DD leads to an incompletely specified function, for which a synthesis scheme similar to the DD-based approach reviewed in Sect. 12.2 can be applied.[2] However, the DD contains $2^{k'}$ times fewer 1-paths (and, hence, contains more 0-paths), i.e., fewer paths have to be considered in order to accomplish the identity structure compared to the established two-stage design flow. Additionally, since the number of 0-paths $p' \in \overline{P}_1$ is larger than the number of 1-paths $p \in P_2$ for most nodes (i.e., $|\overline{P}_1| > |P_2|$) some degree of freedom is introduced. In fact, this is the same degree of freedom which is available when conducting the embedding as described in Sect. 12.1. However, in contrast to the established design flow, this degree of freedom is now exploited during synthesis, because the embedding that suits best (in order to reduce the complexity of the circuit) is implicitly chosen.

More precisely, the degree of freedom allows for choosing the subset of $\overline{P}_1$ containing the 0-paths that require the fewest number of Toffoli gates when swapped with the 1-paths of P_2. Note that the 1-paths through the third edge have to be considered separately, because swapping all 1-paths through the second edge with 0-paths through the first edge does not necessarily swap all 1-paths through the third edge with 0-paths through the fourth edge. In the case that $|\overline{P}_1| = |P_2|$ for the currently considered node, synthesis is conducted as described in Sect. 12.2.

Once the second and third edges are transformed to 0-stubs, the currently considered DD-node does not necessarily establish the desired identity structure— the first or the fourth edge may be a 0-stub as well. In this case, an edge is inserted to accomplish the identity structure (exploiting the degree of freedom that some one-to-one mappings can arbitrarily be defined). In other words, inserting an edge is nothing but adding missing 1-paths in a way that best suits deriving the desired identity structure.

---

[2]Note that there exist approaches for synthesis of incompletely specified functions (see, e.g., [58, 109])—although applicable for rather small functions only. Besides that, the assumption is applied there that the respectively given function can be made reversible by properly assigning the *don't cares* only. The methodology covered in this book is applicable for arbitrary functions and does not rely on such an assumption.

Example 13.3 *Consider the top node of the DD shown in Fig. 13.2 (representing the incomplete function resulting from extending the DD with one additional variable), which contains only four 1-paths. The respective set of paths are $P_1 = \{\overline{x}_1\overline{g}_0, x_1\overline{g}_0\}$, $P_2 = \{x_1\overline{g}_0\}$, $P_3 = \emptyset$, and $P_4 = \{\overline{x}_1\overline{g}_0\}$. The degree of freedom allows for arbitrarily choosing 0-paths from $\overline{P}_1 = \{\overline{x}_1g_0, x_1g_0\}$ that should be swapped with the 1-paths $p \in P_2$. There is only one path $p = x_1\overline{g}_0 \in P_2$. Unfortunately, this path is not contained in $\overline{P}_1$. Therefore, the most similar path $p' \in \overline{P}_1$ is chosen, which is $p' = x_1g_0$. To adjust the paths, gate $TOF(\{x_0^+, x_1\}, g_0)$ is applied, and eventually gate $TOF(\{x_1^+, g_0^+\}, x_0)$ to swap them. The resulting DD is shown on the left-hand side of Fig. 13.3.*

Now, this DD already realizes the desired identity structure for almost all nodes labeled with x_0 and x_1. Only the fourth edge of the right node labeled x_1 is a 0-stub (although a 1-path is required here). However, this is addressed without the need to add any further gates. In fact, the degree of freedom allows simply adding a new 1-path through this edge as shown at the right-hand side of Fig. 13.3. Note that, if embedding had been conducted prior to synthesis, 1-paths would have been set very likely in a fashion which requires additional gates. In contrast, utilizing a one-pass scheme allows setting these 1-paths appropriately without leading to further costs.

The synthesis scheme outlined above only needs to be conducted for DD-nodes representing primary variables. Any additional variables (e.g., g_0 in the example) can be mapped arbitrarily as long as the dependencies discussed in Sect. 12.1 are considered. This, however, is implicitly the case since only reversible gates are applied thus far—eventually realizing a reversible function. Again, this fully exploits the degree of freedom, since a prior embedding step may have realized a mapping of additional variables such as g_0, which would require further gates.

Example 13.3 (Continued) *Considering the current DD as shown at the right-hand side of Fig. 13.3, it can be seen that all nodes labeled by primary variables already realize the desired identity structure. In contrast, the mapping of the additional variable g_0 can arbitrarily be realized (reversibility is guaranteed by the fact that only reversible gates have been employed thus far). Hence, the synthesis process terminates and yields the circuit as shown in Fig. 13.4.*

Fig. 13.3 Handling incompletely specified functions

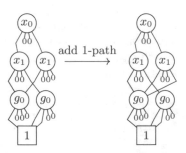

Fig. 13.4 Circuit obtained by
the exact one-pass design

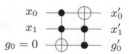

13.1.3 Heuristic Solution

The exact solution presented in the previous section yields circuits with the minimum number of circuit lines. Even though conducted evaluation (summarized later in Sect. 13.1.4) demonstrates the scalability of this approach, one of the drawbacks remains: the DD has to be extended by additionally required variables prior to synthesis—causing an exponential overhead in the worst case (as discussed above). In order to overcome this drawback, a heuristic solution is presented in this section that may require more (although still a bounded number of) additional circuit lines, but instead does not lead to an increase of the function representation.

The main idea is to stay with the originally given function representation (i.e., a DD representing a function matrix without additional variables) and to modify this (non-reversible) function to be synthesized whenever a DD-node is encountered for which the second edge has more 1-paths than the first edge has 0-paths (the problem discussed in Sect. 13.1.1). Since this obviously leads to a circuit that realizes a function different to the desired one, these modifications are stored on the so-called buffer lines and reverted after synthesis terminated.

More precisely, a DD-node with $|\overline{P}_1| \geq |P_2|$ and $|\overline{P}_4| \geq |P_3|$ is transformed to the identity structure as described above.[3] In case the second edge has more 1-paths than the first edge has 0-paths (i.e., $|\overline{P}_1| < |P_2|$) not all 1-paths through the second edge can be moved to the first edge. Then, as many 1-paths through the second edge as possible are swapped with 0-paths through the first edge by applying the respective gates. Afterwards, the remaining 1-paths $p \in P_2$ are swapped with 0-paths $p' \in \overline{P}_4$ to establish the identity structure. This can be conducted by flipping the output of the currently considered variable x_i for these paths (since $P_2 \subseteq \overline{P}_4$). Obviously, this changes the function to be synthesized because bit i is flipped in the output pattern for some inputs. To remember these bit flips on bit i, a so-called buffer line b_{x_i}—initialized with zero—is added onto which all input combinations are stored which are affected by this flip. These input combinations are derived from the respectively considered path, i.e., a Toffoli gate with target line b_{x_i} is applied for each 1-path $p \in P_2$ (the set of control lines contains one variable for each literal in p and in path to the currently considered node, as well as a positive control line x_i^+). An example illustrates the idea:

Example 13.4 *Consider the top node of the DD shown in Fig. 11.2c on Page 99 representing the (non-reversible) function matrix from Table 12.1a on Page 107.*

---

[3]If the number of outputs is larger than the number of inputs, the function matrix describes an incompletely specified function, since it contains column consisting of 0-entries only.

The respective set of paths are $P_1 = \{\overline{x}_1, x_1\}$, $P_2 = \{x_1\}$, $P_3 = \emptyset$, and $P_4 = \{\overline{x}_1\}$. Since $\overline{P}_1 = \emptyset$ does not contain any 0-path that can be swapped with a path from P_2, the synthesis approach described in Sect. 12.2 cannot be applied. Therefore, the function is modified by swapping the 1-path of P_2 with the 0-path of $\overline{P}_4$ to obtain the desired identity structure. To this end, the output x_0 for the path $x_1 \in P_2$ is flipped, resulting in the identity DD. Since this modification has to be restored after synthesis is completed, it is stored on a buffer line b_{x_0} (initialized with zero) by applying Toffoli gate $TOF(\{x_0^+, x_1^+\}, b_{x_0})$ (the first gate in Fig. 13.5).

After synthesis has terminated, the buffer lines are used to revert the made modifications. To this end, for each buffer line b_{x_i} a single gate $TOF(\{b_{x_i}^+\}, x_i)$ is applied, which flips the output back to its intended value.

Example 13.4 (Continued) *Since the modification of the function described above already yields the identity DD, synthesis completes without adding further gates. Finally, the modification made on output x_0 is restored by applying gate $TOF(\{b_{x_0}\}^+, x_0)$, eventually resulting in the circuit shown in Fig. 13.5.*

Although additional circuit lines are added dynamically during synthesis, their number is bound by m, since in worst case each of the m output bits has to be flipped for at least one input combination. Therefore, at most m buffer lines (one for each output bit) are required—yielding a circuit with $n + m$ lines at maximum. Moreover, if the function to be synthesized is reversible, no additional lines are added during synthesis, because the cardinalities of the set of 1-paths and 0-paths always match.

Overall, both approaches presented above realize a one-pass design flow and conduct both, the embedding step and the synthesis step, at the same time. By this, they fully exploit the degree of freedom, as corresponding paths are set so that they perfectly suit the synthesis (ideally, the paths are set so that the desired identity structure is realized with no or significantly fewer gates). Moreover, the heuristic scheme does not even require additional variables for the respective function description and, by this, simplifies synthesis further (at the expense of not guaranteeing a circuit with the minimum number of lines). Table 13.1 summarizes the characteristics of the established design flow, as well as of the presented solutions for one-pass design of reversible circuits. As can be seen, the drawbacks of the established design flow discussed in the beginning of this section are addressed by the presented solutions. The conducted evaluation summarized in the next section confirms the benefits of the new design flow.

Fig. 13.5 Circuit obtained by
the heuristic one-pass design

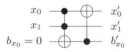

Table 13.1 Characteristics of design flows

| | Established (two-stage) | One-pass design flow | |
		Exact	Heuristic
Exploits full degree of freedom?	✗	✓	✓
No. variables for function representation	$\max(n, m + k)$	$\max(n, m + k)$	$\max(n, m)$
Always yields circuit with min. lines?	✓	✓	✗

13.1.4 Evaluation

In order to evaluate the one-pass design flow, both, the exact solution as well as the heuristic solution (both utilizing the DD-based simulation approach discussed in Sect. 12.2) have been implemented in C++ on top of the DD-package presented in Chap. 7, the BDD-package CUDD [156], and RevKit [149].[4] As benchmarks served the Boolean functions available at RevLib [173] and IWLS93 [100], as well as functions describing adders of different size. The evaluation has been conducted on a machine with 32 GB of memory running Linux 4.4 and the timeout was set to 1 h.

To allow for a fair comparison of the design flows, the same underlying synthesis scheme is utilized. More precisely, the original functional synthesis flow is represented by the embedding provided in Sect. 12.1, which passes the resulting embedded function to the DD-based synthesis approach discussed in Sect. 12.2, and serves as baseline. The presented one-pass design solutions (relying on DD-based synthesis as well) represent this new design flow. This way, a fair comparison between both synthesis flows is conducted.

Table 13.2 presents the obtained results. The first four columns of the table list the name of the benchmark, the number of primary inputs n and primary outputs m, as well as the minimum number of required lines min (these numbers have been determined using the approach described in Sect. 12.1). The remaining columns list the time required for synthesis (columns labeled t) and the T-count of the resulting circuit for the two-stage design flow as well as for the discussed one-pass solutions. Moreover, the number of lines of the resulting circuits l is listed for the heuristic solution (since it does not guarantee minimality). The cases where this number of lines is the actual minimum are highlighted in bold. Note that benchmarks that already represented a reversible function are omitted since, in these cases, all three design flows yield the same circuit.

---

[4]An implementation is available at http://www.jku.at/iic/eda/one_pass_design_of_reversible_circuits.

Table 13.2 Comparison of the design flows using DD-based synthesis

| Benchmark | n | m | min | Two-stage design flow (Chapter 12) | | One-pass design flow | | | | |
| | | | | | | Exact sol. (Sect. 13.1.2) | | Heuristic sol. (Sect. 13.1.3) | | |
				t	T-count	t	T-count	t	l	T-count
alu2	10	6	14	6.55	2,321,575	0.73	429,974	0.28	16	21,225
x2	10	7	16	23.03	5,069,950	0.16	169,143	0.04	17	2885
alu3	10	8	14	12.92	3,202,470	0.3	356,471	0.28	18	28,270
ex1010	10	10	18	17.66	7,206,543	2.84	1,512,214	0.70	20	112,686
dk17	10	11	19	103.36	26,849,673	2.26	1,330,086	0.22	21	19,703
apla	10	12	22	1083.04	57,731,187	1.95	1,298,684	0.24	**22**	38,003
cm152a	11	1	11	0.06	376	0.02	496	0.02	12	352
cm85a	11	3	13	4.18	1,198,960	0.06	56,032	0.05	14	3614
05-adder_col	11	6	12	4.18	2,069,167	1.78	859,062	0.19	17	39,312
add6	12	7	13	10.81	3,970,087	2.89	1,300,663	0.18	**13**	108,756
alu1	12	8	18	70.77	14,952,568	1.21	1,056,493	0.17	20	9024
06-adder_col	13	7	14	43.68	11,478,480	15.6	4,711,190	1.30	20	163,343
co14	14	1	15	0.12	15,815	0.03	19,664	0.02	**15**	4992
alu4	14	8	19	TO	–	76.04	15,335,701	11.48	22	1,034,798
f51m	14	8	19	2980.33	89,529,939	62.63	9,796,031	20.68	22	1,930,263
cu	14	11	25	TO	–	4.91	3,372,623	0.15	**25**	1794
misex3	14	14	28	TO	–	335.78	44,018,885	2.40	**28**	516,361
table3	14	14	28	TO	–	379.92	44,284,390	0.77	**28**	173,284
b12	15	9	22	TO	–	5.94	3,603,999	0.16	24	3573

	n	m	min	t	l	t	l	t	l	l
in0	15	11	25	TO	–	2140.04	62,144,709	11.23	26	594,574
ryy6	16	1	17	87.15	19,132,055	0.26	351,648	0.04	**17**	8368
t481	16	1	17	1202.96	33,195,808	3.79	2,508,064	0.06	**17**	51,751
cmb	16	4	20	TO	–	0.44	982,687	0.07	**20**	752
pcler8	16	5	21	3009.87	83,811,839	7.35	1,499,597	0.10	**21**	789
cm163a	16	13	25	TO	–	37.05	15,216,228	0.32	27	7759
table5	17	15	32	TP	–	TO	–	14.22	**32**	313,539
cm151a	19	9	27	TO	–	TO	–	0.21	28	1015
09-adder_col	19	10	20	TO	–	TO	–	7.20	28	24,546,894
cm150a	21	1	22	TO	–	1.7	164,854	0.23	**22**	640
mux	21	1	22	103.14	14,866,375	1.75	180,790	0.24	**22**	640
10-adder_col	21	11	22	TO	–	TO	–	35.61	31	106,847,758
cordic	23	2	25	TO	–	TO	–	28.36	**25**	1,928,368
11-adder_col	23	12	24	TO	–	TO	–	217.81	34	465,505,742
vg2	25	8	32	TO	–	TO	–	5.25	33	158,328
12-adder_col	25	13	26	TO	–	TO	–	1339.68	37	2,024,140,174
misex2	25	18	42	TO	–	TO	–	0.78	43	4350
frg1	28	3	30	TO	–	605.79	17,961,727	0.54	31	22,446
apex2	39	3	42	TO	–	TO	–	15.79	**42**	402,991
apex1	45	45	89	TO	–	TO	–	434.19	90	5,986,359
e64	65	65	129	TO	–	TO	–	31.25	**129**	687,368

n primary inputs, m primary outputs, min minimum number of required lines, t time required for synthesis in seconds, l number of lines of the resulting circuit. The timeout was set to 1 h.

First, the exact solution (covered in Sect. 13.1.2) is compared to the original design flow composed of two steps. For all benchmarks, a significant improvement is observed in terms of T-count—leading to an improvement of 89.7% on average. Moreover, the one-pass design flow is much more scalable, which allows synthesizing eleven functions that led to a timeout in the baseline. Since the underlying synthesis algorithm of both approaches is equal (namely DD-based synthesis), this improvement is completely attributed to the usage of one-pass synthesis and, hence, the exploitation of the discussed degree of freedom.

Next, the heuristic solution is compared to the exact solution. The heuristic approach is much faster than the exact one since the DD is not enriched by further variables prior to synthesis (in contrast to the two-stage design flow and the exact design flow which are extended to a total of $\max(n, m + k)$ variables, the heuristic solution stays with the originally given $\max(n, m)$ variables). This allows for a more compact representation of the function to be synthesized.

In fact, using the heuristic one-pass synthesis flow allows synthesizing all benchmarks in reasonable time (including the ones which timed out before). Although the number of circuit lines of the resulting circuits is not necessarily the minimum, the conducted evaluation shows that a circuit with the minimum number of lines was generated in 16 out of 40 cases (highlighted bold in column l of Table 13.2). However, the heuristic approach exploited the reduced complexity and generated all results in significantly less run-time. For the benchmarks *cordic*, *apex2*, and *e64* a circuit with the minimum number of lines could be determined within approximately half a minute, whereas the exact approach timed out. As discussed above, this was possible because the heuristic solution does not increase the function representation by additional variables.

Hence, also the second benefit discussed above can fully be exploited here. Besides that, this even yields further improvements with respect to the T-count, since the more variables the function to be synthesized has, the more likely it is that a gate has a larger number of control lines (which causes higher T-count). In fact, another reduction of 98.5% (99.2% if only those benchmarks are considered for which the resulting circuit has a minimum number of lines) compared to the exact solution can be observed—resulting in an average improvement by a factor of 281.29 compared to the original two-stage design flow.

Overall, the conducted evaluation confirms the benefits of the one-pass design flow (which is also applicable to other functional synthesis approaches) over the conventional two-stage design flow. Besides substantial speedups compared to the established design flow, substantial improvements in terms of T-count (a factor of 281.29 on average) were observed—clearly outperforming the currently established functional design flow for reversible circuits where embedding and synthesis are conducted separately.

13.2 Exploiting Coding Techniques in One-Pass Design

This section enriches the presented one-pass design flow with the idea of exploiting coding techniques in order to reduce the number of variables that have to be considered during synthesis.[5] This idea is based on the fact that the output patterns in non-reversible functions are not uniformly distributed—leading to a situation where some patterns require many additional outputs while others require only a few. Hence, several garbage outputs are required only for certain output patters. Avoiding this overhead provides significant potential for improving synthesis. In fact, this section shows that employing a variable-length code allows realizing any non-reversible function with a single ancillary qubit only—allowing conducting synthesis on significantly fewer variables than before. The key idea is to represent frequently occurring output patterns (which require more garbage outputs) with a smaller number of variables. Vice versa, less frequently occurring patterns (which require less garbage outputs) are represented with a larger number of variables. In other words, coding techniques are utilized in order to encode the desired function with a variable-length code in which the length of the code word for an output pattern p_i is indirectly proportional to the number $\mu(p_i)$ of times the pattern occurs. An example illustrates that.

Example 13.5 *Consider the Boolean function shown in Table 13.3a and its distribution of the output patterns as shown in Table 13.3b. Following, e.g., the exact one-pass design flow outlined in Sect. 13.1.2 results in a function with 5 inputs/outputs since the most frequent output pattern $p_1 = 010$ occurs four times and, thus, requires two garbage outputs. However, using a variable-length code as shown in Table 13.3c allows reducing the number of required qubits. There, the most frequent output pattern is encoded by $c(p_1) = 0$. Since this pattern requires*

Table 13.3 Variable-length encoding for one-pass design

(a) Orig. function						(b) Output patterns			(c) Resulting encoding			(d) Encoded function					
x_0	x_1	x_2	y_0	y_1	y_2	i	p_i	$\mu(p_i)$	i	p_i	$c(p_i)$	x_0	x_1	x_2	y_0	y_1	y_2
0	0	0	0	1	0	1	010	4	1	010	0 - -	0	0	0	0	-	-
0	0	1	0	1	0	2	100	2	2	100	1 0 -	0	0	1	0	-	-
0	1	0	1	0	0	3	001	1	3	001	1 1 0	0	1	0	1	0	-
0	1	1	1	0	0	4	011	1	4	011	1 1 1	0	1	1	1	0	-
1	0	0	0	1	1	5	000	0				1	0	0	1	1	1
1	0	1	0	1	0	6	101	0				1	0	1	0	-	-
1	1	0	0	1	0	7	110	0				1	1	0	0	-	-
1	1	1	0	0	1	8	111	0				1	1	1	1	1	0

---

[5]Note that exploiting coding techniques is also possible in the original design flow composed of an embedding and a synthesis step.

two garbage outputs, in total $1 + 2 = 3$ *outputs are required.*[6] *The second most frequent output pattern* $p_2 = 100$ *is encoded by* $c(p_2) = 10$. *Since this pattern occurs only twice, one garbage output is required—again resulting in* $2 + 1 = 3$ *outputs. The patterns* p_3 *and* p_4 *are encoded by* $c(p_3) = 110$ *and* $c(p_4) = 111$, *respectively. Here, no garbage outputs are required. The remaining patterns (* p_5 *to* p_8 *) do not have to be encoded, since they never occur. Overall, this yields an (encoded) reversible function which embeds* f *as shown in Table 13.3d and is composed of a total of 3 inputs/outputs only—two qubits fewer than without using coding.*

Following this idea, at most $n + 1$ qubits—instead of $\max(n, m + \lceil \log_2 \mu(p_1) \rceil)$— are required to embed any non-reversible function with n inputs. Concerning the design of Boolean components contained in quantum algorithms, the encoded outputs can be handled (1) *locally* where decoders are required for each sub-component that again increase the number of qubits to $\max(n, m + \lceil \log_2 \mu(p_1) \rceil)$, or (2) *globally* where subsequent components that are capable of handling encoded inputs allow remaining at $n + 1$ qubits.

Incorporating the idea of utilizing coding techniques into the one-pass design flow introduced above unveils even more potential. In fact, it allows exploiting an even larger degree of freedom since the values of the garbage outputs are basically *don't care* (except the restriction that a reversible function has to be realized)—while still guaranteeing to synthesize a circuit that uses the minimum number of qubits (or even below that minimum if no decoding is required afterwards). This degree of freedom allows for synthesizing circuits with significantly smaller T-count.

The remainder of this section is structured as follows. Section 13.2.1 presents a technique to encode the output patterns such that at maximum one ancillary qubit is required for embedding. This upper bound is proven in Sect. 13.2.2. Afterwards, Sect. 13.2.3 shows how the additional degree of freedom can be exploited in the new one-pass design flow outlined above (again using DD-based synthesis as representative). Section 13.2.4 discusses how encoded output can be handled and provides an efficient method for generating a decoder in case it is required. Finally, the idea of utilizing coding techniques is evaluated in Sect. 13.2.5.

13.2.1   *Determining the Code*

As discussed above, a code is desired that allows for splitting the available outputs into a code word and garbage outputs. Doing this splitting individually for each output pattern allows using short code words (i.e., few primary outputs) for frequently occurring patterns, which require a large number of garbage outputs. By

---

[6]The garbage outputs are represented by a dash, since they represent *don't care* values (as long as it is ensured that the resulting function is reversible).

this, the need for many garbage outputs is compensated by smaller code words. Vice versa, less frequently occurring output patterns are encoded with larger code words, since they do not require many garbage outputs. By definition, a Huffman code [71] can guarantee this. In the following, a small deviation of Huffman encoding is utilized, which is denoted *Pseudo-Huffman* coding.

The code is computed by generating the *Pseudo-Huffman* tree: Starting with terminal nodes—one for each output pattern with $\mu(p_i) > 0$ (no code has to be assigned to output patterns that do not occur)—with attached weights representing the number of respectively required garbage outputs (i.e., $\lceil \log_2 \mu(p_i) \rceil$), the *Pseudo-Huffman* tree is then generated by repeatedly combining the two nodes a and b with the smallest attached weights $w(a)$ and $w(b)$ to a new node c with weight $w(c) = \max(w(a), w(b)) + 1$ until a single node results. The weight of such a node $w(c)$ then gives the number of outputs required to represent all combined output patterns uniquely, i.e., one additional variable is required (aside from $\max(w(a), w(b))$) to distinguish between a and b.

Example 13.6 *Consider again the distribution of the output patterns as shown in Table 13.3b. Determining the Pseudo-Huffman code starts with the nodes v_1, v_2, v_3, and v_4—one for each output pattern p_i with $\mu(p_i) > 0$. These nodes are shown at the bottom of Fig. 13.6. The weights are drawn inside the respective nodes. The weight of node v_1 is $w_1 = k_1 = 2$, because output pattern $p_1 = 010$ requires two garbage outputs. The weights of the nodes representing p_2, p_3, and p_4 are 1, 0, and 0, respectively. In the first step, the nodes v_3 and v_4 (both have weight 0) are combined. The resulting node v_5 has a weight of $w_5 = \max(0, 0) + 1 = 1$. Next, the two nodes with weight 1 (i.e., v_2 and v_5) are combined. The resulting node v_6 has a weight of $w_6 = \max(1, 1) + 1 = 2$. Finally, the two remaining nodes are combined to a new node v_7 with weight $w_7 = \max(2, 2) + 1 = 3$—eventually resulting in the tree shown in Fig. 13.6.*

After generating the Pseudo-Huffman tree, the overall number of variables that are required to realize the encoded function is given by the weight of the root node of the tree. The resulting code is inherently given by the structure of the Pseudo-Huffman tree. In fact, each path from the root node to a leaf node represents a code word, where taking the left (right) edge implies a 0 (1).

Fig. 13.6 Huffman tree for the function from Table 13.3a

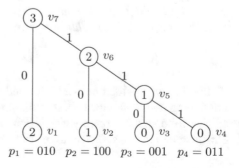

Example 13.6 (Continued) *Since the root node has a weight of 3, three variables are required to realize the encoded function (without encoding, $\max(3, 3 + 2) = 5$ variables would be required). The path from the root node to the leaf node v_2 (which represents output pattern p_2) traverses the right edge of the root node v_7 as well as the left edge of v_6. Consequently, $c(p_2) = 10$ encodes $p_2 = 100$. Since v_2 has weight $w_2 = 1$, one output is used as garbage output in this case. Accordingly, code words for all other output patterns are determined—eventually resulting in the code shown in Table 13.3c. Dashes again represent* don't cares.

Note that the number of output patterns that may occur at least once increases exponentially with the total number of inputs—a significant threat to the coding approach with respect to scalability. However, the conducted evaluation (later summarized in Sect. 13.2.5) shows that, for most functions, the number of output patterns p_i with $\mu(p_i) > 0$ is feasible (the vast majority of output patterns never occur and, hence, do not need to be considered in the code).

13.2.2   Proving an Upper Bound of $n + 1$ Qubits

This section proves that encoding the output patterns of an n-input function as shown above results in a coded function requiring at most $n + 1$ variables, i.e., that it is sufficient to add only one qubit to embed any non-reversible Boolean function into a (encoded) reversible one. Moreover, it is shown precisely in which cases this additional qubit is required and in which not. To this end, the Pseudo-Huffman tree utilized to determine the encoding is formally defined.

Definition 13.1 (Pseudo-Huffman Tree) *Let $G = (V, E)$ be a connected, arborescence (a directed rooted tree) composed of a set of nodes $V = \{v_1, v_2, \ldots, v_{|V|}\}$ and a set of edges $E \subset V \times V$, and let $w \colon V \to \mathbb{N}_0$ be a labeling of the graph nodes in terms of non-negative weights. Moreover, let $T = \{t \in V \mid \forall v \in V \colon (v, t) \notin E\} \subseteq V$ denote the set of all terminal nodes. Then, $PH = (G, w)$ is called a* Pseudo-Huffman tree, *if, and only if,*

1. *each internal node $v \in V \setminus T$ has exactly two children $a, b \in V$ and $w(v) = \max(w(a), w(b)) + 1$ and*
2. *for any two different internal nodes $v_1, v_2 \in V \setminus T$ with children a_1, b_1 and a_2, b_2, respectively, it holds that $w(a_1) \leq w(b_1)$ implies*

$$\big(w(a_2), w(b_2) \leq w(a_1)\big) \vee \big(w(a_2), w(b_2) \geq w(b_1)\big).$$

In other words, the tree can be formed from the terminal nodes by successively combining nodes with the lowest available weights as described in Sect. 13.2.1.

The following Theorem yields a condition on the terminal nodes of a Pseudo-Huffman tree that is sufficient to restrict the weight of the tree's root node.

Theorem 13.1 *Let $PH = ((V, E), w)$ be a Pseudo-Huffman tree. If there exists an assignment s_v for each terminal node $v \in T = \{t \in V \mid \forall v \in V : (v, t) \notin E\}$ such that $2^{w(v)} \geq s_v > 2^{w(v)-1}$ (where $w(v)$ denotes the weight of node v) and $\sum_{v \in T} s_v = 2^n$, then the weight $w(v_r)$ of the root node v_r of the tree is either n or $n + 1$.*

Proof Replace all weights using the rule $w \mapsto 2^w$. Then the rule for computing the weight of a new node changes from $\max(w(a), w(b)) + 1$ to $2 \cdot \max(w(a), w(b))$. Accordingly, all weights in the tree will be a potency of 2.

The proof argues about the weights of the nodes when constructing a Pseudo-Huffman tree. To this end, consider the set of all nodes V_i of the tree-under-construction that are the root nodes of the already connected components after step i of the algorithm. Let $w_{total}^i = \sum_{v \in V_i} w(v)$ denote the sum of the weights over all these nodes.

At each step i of the algorithm, two nodes $a, b \in V_i$ with minimal weight are chosen and joined to a new node c such that $V_{i+1} = \{c\} \cup V_i \setminus \{a, b\}$. There are two cases:

1. both nodes a and b have the same weight 2^k. Then, they are replaced by a node with weight 2^{k+1} such that $w_{total}^{i+1} = w_{total}^i$ (see Fig. 13.7a), i.e., the sum of the weights over the root nodes remains constant.
2. one node—assume without loss of generality a—has weight $w(a) = 2^k$ and the other node (b) has weight $w(b) = 2^{k-l}$ for some $k \geq l > 0$. Then, they are replaced by a node c with weight $w(c) = 2^{k+1}$ (see Fig. 13.7b).

   Since always the nodes with minimal weight are taken, there might not be any other node $d \in V_i$ with $w(d) < 2^k$ as this node would have a higher priority to be joined with b. Thus, all nodes in V_i aside from b have a weight that—by construction—is a potency of 2 that is greater than or equal to 2^k. Consequently, after joining a and b, w_{total}^i is increased to a number w_{total}^{i+1} that is divisible by 2^k. More precisely, it is increased by

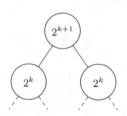

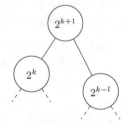

(a) Joining nodes with equal weights   (b) Joining nodes with different weights

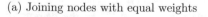

Fig. 13.7 Joining nodes in the construction of the PH-tree

$$w(c) - w(a) - w(b) = 2^{k+1} - 2^k - 2^{k-l}$$
$$= 2^k - 2^{k-l}$$
$$< 2^k,$$

such that w_{total}^{i+1} is the *smallest* number that is greater than w_{total}^i and divisible by 2^k.

Clearly, this case happens at most once for each $k > 0$, since afterwards there is no more node in V_{i+1} with a weight less than 2^k and all nodes that will be added to $V_{j>i}$ have higher weights.

By assumption of the Theorem, $2^n = \sum_{v \in T} s_v \le w_{total}^0$ and $w_{total}^0 < 2^{n+1}$, initially. Thus, at some point $final$ the case is reached that all nodes in V_{final} have a weight greater than or equal to 2^n such that w_{total}^{final} is divisible by 2^n. Since 2^{n+1} is divisible by all potencies 2^k for $k = 0, \ldots, n$, w_{total}^{final} will never exceed 2^{n+1}, as w_{total}^i is always increased to the smallest larger number divisible by 2^k for a $k \in \{1, \ldots, n\}$. Consequently, at most two nodes in V_{final} with a weight of 2^n result, implying that the root node of the resulting tree either is the single weight-2^n node or the single weight-2^{n+1} node constructed from the two weight-2^n nodes. Thus, the root node of the original Pseudo-Huffman tree has weight n or $n + 1$ as desired.

This result is interpreted in the setting of coded Boolean functions as follows. Consider a Boolean function $f : \mathbb{B}^n \to \mathbb{B}^m$ to be encoded and assume that a Pseudo-Huffman tree with $|T| = |\{p_i \in \mathbb{B}^m \mid \mu(p_i) > 0\}|$ terminal nodes is constructed (which is always possible), where each terminal node $v \in T$ uniquely corresponds to one output pattern p_i and has assigned $s_v = \mu(p_i)$ (thus, having a weight $w(v) = \lceil \log_2 \mu(p_i) \rceil$). As this assignment clearly satisfies the conditions of Theorem 13.1, the height of this tree is either n or $n + 1$. Hence, there exists a coding (which is inherently given by the constructed tree) that requires at most one additional qubit when realizing f in quantum logic.

Moreover, the cases in which this additional qubit is required are precisely determined. In fact, the additional qubit is required whenever there exists an output pattern p_i where $\mu(p_i) > 0$ is not a power of two.

Corollary 13.1 *The root node of a Pseudo-Huffman tree satisfying the same assumptions as in Theorem 13.1 has weight n if, and only if, $\sum_{v \in T} 2^{w(v)} = 2^n$.*

Proof Given that $\sum_{v \in T} 2^{w(v)} = 2^n$ Theorem 1 can be applied by using the assignment $s_v = 2^{w(v)}$ for all $v \in T$. Following the argumentation in the proof of Theorem 1, the root node of the Pseudo-Huffman tree has weight n if w_{total}^i does not exceed 2^n at any time, i.e., if the second case (which increases w_{total}^i) does not occur a single time. This is clearly the case if $w_{total}^0 = 2^n$ in the beginning.

Conversely, if $\sum_{v \in T} 2^{w(v)} \neq 2^n$, it holds that $w^0_{total} > 2^n$ in the beginning, such that $w^{final}_{total} = 2^{n+1}$ in the end.

13.2.3 Exploiting the Encoding During Synthesis

In contrast to the established design process, the Pseudo-Huffman code yields a function to be synthesized in which single outputs may be used as both, actual primary outputs as well as garbage outputs—dependent on the output pattern. Besides the fact that this allows for performing the actual synthesis with significantly fewer variables, it inherently motivates a synthesis scheme in which a huge degree of freedom can be exploited. As in the one-pass design flow introduced in Sect. 13.1, the *don't care* values can be arbitrarily assigned since their value does not matter as long as a reversible function results. Since this is an inherently given property of reversible circuits, their value is *don't care*.

Example 13.7 *Applying the determined Pseudo-Huffman code (cf. Table 13.3c) to the output patterns of the function shown in Table 13.3a yields the encoded function shown in Table 13.3d. The corresponding permutation matrix is depicted in Fig. 13.8a. In the original function, the input combinations 000, 001, 101, and 110 are all mapped to the output patterns $p_1 = 010$. Since this output pattern is now encoded by $0 - -$, there are four possibilities for each of the input combinations where to locate the corresponding 1-entry in the permutation matrix (the only requirement is that the 1-entries must be in different rows to guarantee reversibility). This degree of freedom is indicated by a blue rectangle in Fig. 13.8. A similar degree of freedom (indicated by a red rectangle in Fig. 13.8) occurs for output pattern 100, which is encoded by $1\,0\,-$.*

To exploit the *don't care* values as much as possible, one variable is transformed after the other to the identity—starting at the most significant variable. Since this basically describes the general idea of DD-based synthesis discussed in Sect. 12.2, it is modified to exploit *don't care* values whenever possible.

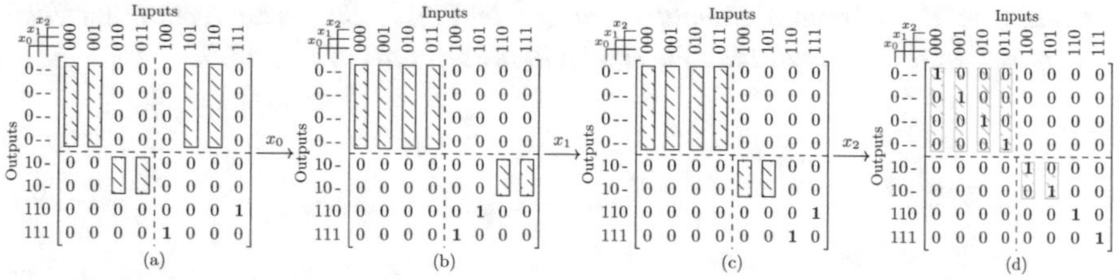

Fig. 13.8 Synthesis of the encoded function

Example 13.7 (Continued) *To establish the identity for the most significant variable x_0 in the permutation matrix shown in Fig. 13.8a, the columns 101 and 110 are swapped with columns 010 and 011, respectively. This can, e.g., be achieved by applying the Toffoli gates $TOF(\{x_2^+, x_0^+\}, x_1)$ and $TOF(\{x_1^+\}, x_0)$. The first gate exchanges columns 101 and 111 (by inverting x_1), whereas the second Toffoli gate simultaneously swaps columns 110 and 111 with columns 010 and 011. The resulting permutation matrix is shown in Fig. 13.8b.*

In the considered DD-based synthesis method, the number of control lines that are added to the gates to transform the corresponding sub-matrix to the identity increases with recursion depth. Although the number of these additional control lines can be reduced by exploiting redundancies in the permutation matrix (cf. Sect. 12.2.2), this usually yields rather expensive circuits (as reviewed in Sect. 2.3, the respective costs depend on the number of control lines in each gate). However, by exploiting the available degree of freedom (provided by the *don't care* values), this overhead is significantly reduced: Assume without the loss of generality that the most significant variable x_0 has just been transformed to the identity and that all other variables $x_1 \ldots x_{n-1}$ are *don't care* in case $x_0 = 0$ (i.e., the top left quadrant is *don't care*). This is the case when the most frequent output pattern p_1 is encoded with $c(p_1) = 0$. Since all values in the top left quadrant are *don't care*, it does not matter which gates are applied to that quadrant. The only constraint is that a reversible function results, which is inherently given since reversible gates are applied only. Consequently, it is not necessary to care whether the gates that are applied to transform the bottom right quadrant also affect the top left quadrant. Therefore, no additional control line has to be added.

Example 13.7 (Continued) *Since the left upper quadrant is composed of* don't *care values only, no positive control line x_0 has to be added when applying the algorithm recursively to the bottom right quadrant. Transforming the most significant variable of this sub-matrix (i.e., x_1) to the identity requires applying a Toffoli gate $TOF(\emptyset, x_1)$. The resulting permutation matrix is shown in Fig. 13.8c. Since again the top left quadrant is composed of* don't *cares, no additional control lines are required when applying the algorithm recursively to the 2×2 sub-matrix in the bottom right corner of the permutation matrix. Finally, the Toffoli gate $TOF(\emptyset, x_0)$ is required to transform this sub-matrix to the identity—eventually resulting in the permutation matrix shown in Fig. 13.8d.[7] The resulting circuit (composed of three circuit lines) is shown in the left part of Fig. 13.9.*

---

[7]The *don't care* values are assigned along the main diagonal. Therefore, no further gates are required.

Fig. 13.9 Resulting circuit

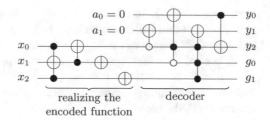

realizing the
encoded function

decoder

13.2.4 Handling Encoded Outputs

As mentioned above, coding techniques allow for a reduction of the variables that are considered during synthesis. However, since this realizes a reversible function that is different to the desired one, the code outputs must be handled. Concerning the design of Boolean components contained in quantum algorithms, two different ways are possible:

- On the one hand, one can apply the coding technique *locally* on each and every sub-component and use decoders (after each sub-component) to translate the encoded results to the original ones which are then used as inputs of the subsequent components. This essentially reduces the complexity of synthesis for the individual sub-components (since a smaller number of qubits needs to be considered). While this offers an improvement of synthesis itself, the total number of additional qubits does not change (due to the decoders).
- On the other hand, one can apply the coding technique *globally* such that the encoded outputs of one sub-component are directly used as input for subsequent components and a single decoder at the end translates the final results to the desired ones. This approach significantly reduces the number of extra qubits required during the computation of the oracle's sub-components such that the total number of extra qubits is likely to stay close to the theoretical minimum given by the oracle's overall functionality. On the downside, a re-design of the sub-components might be required in order to work with encoded values.

Assuming that the subsequent Boolean components are not capable of handling coded outputs, a decoder is required, which recalculates the original outputs. This decoder adds the variables that were saved during the actual synthesis—yielding a circuit with a total number of max $(n, m + \lceil \log_2 \mu(p_1) \rceil)$ inputs/outputs.

A large portion of the decoder can be realized very efficiently: Assume that r additional variables are required such that the overall number of variables is again $\max(n, m + \lceil \log_2 \mu(p_1) \rceil)$. Then, r ancillary lines $a_0, \ldots a_{r-1}$ (initialized with zero) are added to the circuit that realizes the encoded function. These r lines are used to decode the r most significant bits of the original output patterns (i.e., $y_0, \ldots, y_{r-1}$). To this end, all codewords $c(p_i)$ are traversed. If the codeword encodes an output pattern p_i with $y_j = 1$ $(0 \le j < r)$, the value of circuit line a_j is inverted for that codeword. To this end, a Toffoli gate with target line a_j and a set of control lines that represent codeword $c(p_i)$ is applied. More precisely, the set of control lines contains a positive (negative) control line for each 0 (1) in $c(p_i)$. This procedure allows realizing r primary outputs very efficiently. An example demonstrates the idea.

Example 13.8 *Recall the non-reversible function shown in Table 13.3a and the resulting Pseudo-Huffman code for the output patterns shown in Table 13.3c. The left part of Fig. 13.9 shows the realization of the encoded function (cf. Table 13.3d). Since the original function requires* $\max(3, 3 + 2) = 5$ *variables,* $r = 2$ *further circuit lines* (a_0 *and* a_1) *are added to the circuit. These lines decode the primary outputs* y_0 *and* y_1. *As discussed above, the codewords shown in Table 13.3c are traversed. Codeword* $c(p_1) = 0$ *encodes output patterns* $p_1 = y_0 y_1 y_2 = 010$. *To decode* $y_1 = 1$ *for output pattern* p_1, *a Toffoli gate* $TOF(\{x_0^-\}, a_1)$ *is added to the circuit. Analogously, the gates* $TOF(\{x_0^+, x_1^-\}, a_0)$ *and* $TOF(\{x_0^+, x_1^+, x_2^+\}, a_1)$ *are applied to decode codewords* $c(p_2) = 10$ *and* $c(p_4) = 111$ *(cf. right part of Fig. 13.9).*

Note that this procedure cannot be applied to the remaining $m - r$ the primary outputs, because these outputs have to be decoded on lines that are not initialized with constant zero. Hence, these remaining primary outputs have to be decoded using any functional synthesis algorithm (no embedding is required since the function is already reversible). However, since this is necessary only for a small number of primary outputs, it hardly affects the scalability of the overall synthesis result. This is also confirmed by the conducted evaluation.

Example 13.8 (Continued) *Finally, the least significant primary output* y_2 *is adjusted. Note that for all code words except for* $c(p_2) = 10$ *the most significant bit of the code word is equal to* y_2 *of the original output pattern. Adding a Toffoli gate* $TOF(\{a_0^+\}, x_0)$ *establishes the desired mapping also for this case—eventually resulting in the circuit shown in Fig. 13.9. The first four gates of the circuit realize the encoded function and, thus, operate on three lines only. The remaining gates realize the decoder and, therefore, may operate on all five circuit lines.*

13.2.5   Evaluation

This section evaluates the idea of utilizing coding techniques in the one-pass design flow. To this end, the exact one-pass design flow introduced in Sect. 13.1 (utilizing DD-based synthesis) has been enriched with the Pseudo-Huffman code. As benchmarks served again functions available at RevLib [173] and IWLS93 [100], as well as functions describing adders of different size.[8] The evaluation has been conducted on a 4 GHz machine with 32 GB of memory. The timeout was set to 1 h.

Table 13.4 shows the results of the conducted evaluation. The first four columns list the name of the benchmark, its number of primary inputs n and primary outputs m, as well as the minimum number of variables (circuit lines) min that are required to realize the function as reversible circuit. The next two columns list the run-time t as well as the T-count of the synthesized circuits when following the exact one-pass design flow presented in Sect. 13.1.2. Eventually, the remaining columns list the run-time as well as the T-count of the synthesized circuit when additionally considering coding techniques—once for the case that coding is handled globally (resulting in a circuit with n or $n + 1$ lines) and once for the case that coding is handled locally (requiring a decoder that increases the number of lines to min). For the global case, the number of required lines is also listed in the column l.

Although Table 13.4 shows that utilizing coding techniques slightly reduces scalability of the one-pass design flow, several benefits are observed. In fact, assuming that the coding scheme can be applied globally, the number or circuit lines to realize the considered functions reduces significantly. As proven in Sect. 13.2.2, at most one additional qubit is required to realize the considered function. Table 13.4 shows that this allows realizing 17 out of 22 benchmarks (in which the coded function could be synthesized within the timeout) with fewer qubits than what is usually considered the minimum. On average, the lower bound on the number of required qubits (without using coding) is undercut by 19.3%.

Moreover, the costs of the resulting circuits significantly reduce. In the majority of the cases, the utilizing coding allows for significantly smaller circuits. Only in rare cases circuits are obtained that are more complex than those generated by the baseline (i.e., the one-pass-design flow without utilizing coding techniques). Considering that coding techniques are applied globally (which yields circuits composed of fewer circuit lines), a reduction of the T-count by 86.2% (i.e., a factor of more than 7.27) is observed on average (although benchmark *cm152a* yields a circuit with significantly larger T-count). If the coding is applied locally, a decoder is required—again increasing the number of circuit lines to min. However, a reduction of the T-count of 69.6% (i.e., a factor of 3.29) is still achieved on average. These significant reductions in T-count are explained by the fact that significantly fewer variables were considered during synthesis in many cases and that a large degree of freedom is exploited—leading to gates with fewer control lines.

---

[8]Benchmarks that already describe a reversible function are neglected, since in these cases coding is not beneficial.

Table 13.4 Evaluation of the exact one-pass design flow

Name	n	m	min	Without coding		Coding applied				
						l	Globally		Locally	
				t	T-count		t	T-count	t	T-count
alu2	10	6	14	0.73	429,974	11	0.61	107,502	1.08	268,516
x2	10	7	16	0.16	169,143	11	0.02	5711	0.03	14,933
alu3	10	8	14	0.30	356,471	11	0.69	177,588	1.13	416,370
ex1010	10	10	18	2.84	1,512,214	–	TO	–	–	–
dk17	10	11	19	2.26	1,330,086	11	0.13	47,393	0.35	166,862
apla	10	12	22	1.95	1,298,684	11	0.20	36 941	0.25	41,189
cm152a	11	1	11	0.02	496	11	0.05	62,160	0.06	124,224
cm85a	11	3	13	0.06	56,032	12	0.16	41,773	0.27	90,677
05-adder_col	11	6	12	1.77	859,062	12	2.80	386,948	4.73	1,514,850
add6	12	7	13	2.89	1,300,663	13	8.16	1,025,188	18.17	4,049,385
alu1	12	8	18	1.21	1,056,493	–	TO	–	–	–
06-adder_col	13	7	14	15.60	4,711,190	14	19.19	2,252,153	36.64	9,074,567
co14	14	1	15	0.03	19,664	15	0.01	1344	0.02	2688
alu4	14	8	19	76.04	15,335,701	15	41.86	3,826,625	56.70	8,069,279
i51m	14	8	19	62.63	9,796,031	15	58.56	4,734,979	68.54	9,823,559

	n	m								
cu	14	11	25	4.91	3,372,623	15	0.08	11,253	0.37	37,955
misex3	14	14	28	335.78	44,018,885	–	TO	–	–	–
misex3c	14	14	28	359.00	42,869,502	–	TO	–	–	–
b12	15	9	22	5.94	3,603,999	16	24.86	3,100,449	84.70	10,580,574
in0	15	11	25	2140.04	62,144,709	16	32.43	4,259,980	45.90	6,503,564
ryy6	16	1	17	0.26	351,648	17	0.03	2656	0.04	3728
t481	16	1	17	3.79	2,508,064	17	0.01	256	0.02	256
cmb	16	4	20	0.44	982,687	17	0.01	3544	0.02	6116
pcler8	16	5	21	7.35	1,499,597	17	38.28	3,073,234	38.70	4,032,999
cm163a	16	13	25	37.05	15,216,228	17	1236.07	19,856,124	1257.87	41,366,180
cm150a	21	1	22	1.70	164,854	–	TO	–	–	–
mux	21	1	22	1.75	180,790	–	TO	–	–	–
cordic	23	2	25	TO	–	–	17.62	4,389,200	133.58	8,667,591
frg1	28	3	30	605.79	17,961,727	–	TO	–	–	–

n primary inputs, m primary outputs, min minimum number of required lines, t time required for synthesis in seconds, l number of lines of the resulting circuit. The timeout was set to 1 h

Chapter 14
Summary

Abstract This chapter summarizes Part III of this book by reflecting the presented synthesis methods as well as the gained improvements compared to the state of the art.

Keywords Quantum circuits · Boolean complements · Reversible circuits · Decision diagrams · Qubits · Embedding

The design of Boolean components for quantum logic often requires—due to the complexity of synthesis—to decompose the desired functionality into sub-components. Although the overall functionality of the Boolean component is inherently reversible, these sub-functions may not. Realizing these non-reversible sub-functions in quantum logic requires adding additional variables in order to embed them into a reversible function. Since each variable of a function is eventually represented by physical entities on a quantum computer, i.e., qubits, their number shall be kept as small as possible—a non-trivial task.

This part of the book provided methods to enhance the current state of the art for designing Boolean components for quantum computations. This includes improvements to both steps occurring in the functional design flow of sub-components (i.e., embedding and synthesis). Moreover, an entirely new *one-pass* design flow has been presented, which combines embedding and synthesis and, thus, increases the available degree of freedom that can be exploited while reducing the complexity of the function to be realized. Eventually, this new design flow has been extended by a variable-length encoding of the output patterns that provably reduces the number of additionally required qubits/variables to one—allowing to reduce the number of required qubits below what has been considered the minimum thus far. Even if subsequent Boolean components cannot handle encoded inputs (thus, requiring a decoder afterwards that increases the number of variables to the "minimum" again), using this idea allows reducing the number of variables that have to be considered during synthesis significantly and exploiting a much larger degree of freedom.

© Springer Nature Switzerland AG 2020
A. Zulehner, R. Wille, *Introducing Design Automation for Quantum Computing*,
https://doi.org/10.1007/978-3-030-41753-6_14

Overall, these contributions allow reducing the costs (by means of T-count) of the generated circuits by several orders of magnitude in many cases while keeping the number of required qubits at a minimum (or even below that if it is assumed that subsequent components are capable of handling coded inputs). Moreover, the scalability of designing Boolean components with a minimal number of variables has been increased significantly.

Part IV
Mapping Quantum Circuits to NISQ Devices

Chapter 15
Overview

Abstract This chapter sets the context for the quantum-circuit mapping algorithms (for NISQ devices) presented in this part of the book. More precisely, the considered design task is reviewed—mapping the logical qubits of the quantum circuit to the physical ones of the quantum device while satisfying all coupling-constraints given by its architecture. Determining a minimal solution has been proven to be an $\mathcal{NP}$-complete problem, which also settles the requirement for heuristic approaches to cope with larger circuits and target architectures.

Keywords Quantum circuits · Quantum gates · Mapping · SWAP gates · IBM Q · Coupling map · Qubits

While physical realizations of quantum computers have initially been considered an "academic dream," big players like *IBM*, *Google*, *Microsoft*, and *Intel* as well as specialized startups like *Rigetti* and *IonQ* recently entered the field—leading to a situation where quantum computers are reaching feasibility and practical relevance (cf. Sect. 2.2). However, in order to use these currently developed *Noisy Intermediate-Scale Quantum* (NISQ) devices, the quantum algorithm to be executed has to be properly compiled to these devices such that their underlying physical constraints are satisfied. This part of the book focuses on the mapping step of this compilation process. To this end, it is assumed that the considered quantum algorithm has already been translated into a quantum circuit composed of multiple-controlled one-qubit gates. For the "quantum part" of the algorithm, this is often inherently given (e.g., by using components for which such translations are known) or done by hand (even though some automated methods exist [116, 117, 119]). For the "Boolean part" of the algorithm, a gate-level description is often gained by *reversible circuit synthesis* (cf. Part III of this book).

Mapping quantum circuits to NISQ devices requires the consideration two aspects. First, the occurring gates have to be decomposed into elementary operations provided by the target device—usually a single two-qubit gate as well as a broader variety of one-qubit gates to gain a universal gate set. Second, the *logical* qubits of the quantum circuit have to be mapped to the *physical* qubits of the target device while satisfying the so-called *coupling-constraints* given by the respective device. Since not all physical qubits are coupled directly with each other (due to missing physical connections), two-qubit gates can only be applied to selected pairs of physical qubits. Since it is usually not possible to determine a mapping such that all coupling-constraints are satisfied throughout the whole circuit, the mapping has to change dynamically. This is achieved by inserting additional gates, e.g., realizing SWAP operations, in order to "move" the logical qubits to other physical ones.

While there exist several methods to address the first issue, i.e., how to efficiently decompose multiple-controlled one-qubit gates into elementary operations (see [6, 11, 99, 106, 139, 180]), there is hardly any work on how to efficiently satisfy the coupling-constraints of real devices. Although there are similarities with recent work on nearest-neighbor optimization of quantum circuits as proposed in [131, 136, 137, 175–177, 185], they are not applicable since simplistic architectures with one-dimensional or two-dimensional layouts are assumed which have a fixed coupling (all adjacent qubits are coupled) that does not allow modeling all current NISQ devices.

This part of the book investigates the mapping of the logical qubit of a quantum circuit to the physical ones of a NISQ device from a design automation perspective. Thereby, *IBM Q devices* are considered as representatives for NISQ devices to discuss the occurring challenges in detail, as well as to describe the presented solutions. IBM's approach has been chosen, since it provides the first publicly available quantum devices (available since 2017) that can be accessed by everyone through cloud access. Moreover, their coupling-constraints are described more flexibly than those of other companies—allowing to map their coupling-constraints to IBM's model as well. Hence, the developed methods presented in the following chapters of this part of the book are directly applicable to all currently developed NISQ devices.

The following mapping algorithms are presented: First, an *exact* approach is presented (based on [170]) that determines a mapping with minimal or close-to-minimal overhead by using powerful reasoning engines (cf. Sect. 3.3)— providing a lower bound for evaluating heuristic approaches. But since the mapping problem to be solved is $\mathcal{NP}$-complete [18, 144], exact approaches have limited scalability. Hence, a heuristic approach based on informed search (cf. Sect. 3.2) is additionally provided (based on [189, 190]) that significantly outperforms IBM's own solution provided by means of the Python SDK *Qiskit* [33]. Eventually, a dedicated approach for a certain set of random circuits (utilized for validating quantum computers [34]) is presented (based on [199]) that has been declared winner of the *Qiskit Developer Challenge* [80] since it performed significantly better than the other submissions (according to IBM).

To provide a solid basis for the following chapters, Sect. 15.1 gives a brief overview of available efficient methods that conduct the decomposition step. Afterwards, Sect. 15.2 discusses the challenge of satisfying coupling-constraints given by the device when mapping the logical qubits of a quantum circuit to the physical ones of the target hardware—an $\mathcal{NP}$-complete task. Eventually, Sect. 15.3 briefly outlines the considered approaches to tackle this problem in an efficient fashion.

15.1 Decomposing into Elementary Operations

As discussed above, the multiple-controlled one-qubit gates of the quantum circuit to be mapped have to be decomposed into elementary operations provided by the target hardware. To this end, IBM has developed the quantum assembly language OpenQASM [35] that supports specification of quantum circuits. Besides elementary gates, the language allows the definition of complex gates that are composed from the elementary operations CNOT and $U(\theta, \phi, \lambda)$. These gates can then be nested to define gates that are even more complex. Consequently, as long as a decomposition of the gates used in a description of the desired quantum functionality is provided by the circuit designer, the nested structures are just flattened during the mapping process.

In case the desired quantum functionality is not provided in OpenQASM, decomposition approaches such as those proposed in [6, 11, 99, 106, 139, 180] can be applied which determine (e.g., depth optimal) realizations of quantum functionality for specific libraries like *Clifford+T* [19] or *NCV* [11]. Gates from these libraries can then be easily translated into sequences of CNOT and $U(\theta, \phi, \lambda)$ gates.[1] Several automated decomposition methods are available in Quipper (a functional programming language for quantum computing [57]), the ScaffCC compiler for the Scaffold language [4, 82], RevKit [149], and IBM's SDK Qiskit [33].

Example 15.1 *Recall Fig. 2.5 on Page 20, which shows the decomposition of a Toffoli gate into the Clifford+T library (using the approach proposed in [6]). As can be seen, the Toffoli gate is decomposed into 7 CNOTs, 2 H gates, 4 T gates, and 3 $T^\dagger$ gates. While the CNOT gate is an elementary operation on the IBM Q devices, the one-qubit operations are realized as $H = U(\pi/2, 0, \pi)$, $T = U(0, 0, \pi/4)$, and $T^\dagger = U(0, 0, -\pi/4)$, respectively.*[2]

Hence, decomposing the desired quantum functionality to the elementary gate library is already well covered by corresponding related work. Unfortunately, this is not the case for the mapping task, which is discussed next.

---

[1]For example, arbitrary one-qubit gates can be translated into a $U(\theta, \phi, \lambda)$ gate by conducting an Euler decomposition.

[2]Note that the realization by means of U gates might deviate from the matrices specified in Sect. 2.1 by a global phase.

15.2 Satisfying Coupling-Constraints

To satisfy the coupling-constraints, one has to map the n logical qubits $q_0, q_1, \ldots, q_{n-1}$ of the decomposed circuit to the $m \geq n$ physical qubits $Q_0, Q_1, \ldots, Q_{m-1}$ of the considered quantum device such that all coupling-constraints given by the corresponding coupling map (cf. Definition 2.3 on Page 16) are satisfied. Unfortunately, it is usually not possible to find a mapping such that the coupling-constraints are satisfied throughout the whole circuit (this is already impossible if the number of other qubits, a logical qubit interacts with, is larger than the maximal degree of the coupling map). More precisely, the following problems—using $CNOT(q_c, q_t)$ to describe a CNOT gate with control qubit q_c and target qubit q_t, and CM to describe the edges of the device's coupling map—may occur:

- A CNOT gate $CNOT(q_c, q_t)$ shall be applied while q_c and q_t are mapped to physical qubits Q_i and Q_j, respectively, and $(Q_i, Q_j) \notin CM$ as well as $(Q_j, Q_i) \notin CM$.
- A CNOT gate $CNOT(q_c, q_t)$ shall be applied while q_c and q_t are mapped to physical qubits Q_i and Q_j, respectively, and $(Q_i, Q_j) \notin CM$ while $(Q_j, Q_i) \in CM$

To overcome these problems, one strategy is to insert additional gates into the circuit to be mapped. More precisely, to overcome the first issue, one can insert the so-called SWAP operations into the circuit that exchange the state of two physical qubits and, by this, "move" around the logical ones—changing the mapping dynamically.

Example 15.2 *Figure 15.1 shows the effect of a SWAP gate as well as its decomposition into elementary gates supported by the IBM Q devices. Assume that the logical qubits q_0 and q_1 are initially mapped to the physical ones Q_0 and Q_1, respectively (indicated by $\rightarrow$). Then, by applying a SWAP gate, the states of Q_0 and Q_1 are exchanged—eventually yielding a mapping where q_0 and q_1 are mapped to Q_1 and Q_0, respectively.*

The second issue may also be solved by inserting SWAP operations. However, it is cheaper (fewer overhead is generated) to insert four Hadamard operations (labeled by H) as they switch the direction of the CNOT gate (i.e., they change the target and the control qubit). This can also be observed in Fig. 15.1, where $H = U(\pi/2, 0, \pi)$ gates switch the direction of the middle CNOT in order to satisfy all coupling-constraints given by the coupling map (assuming that only CNOTs with control qubit Q_1 and target qubit Q_0 are possible).

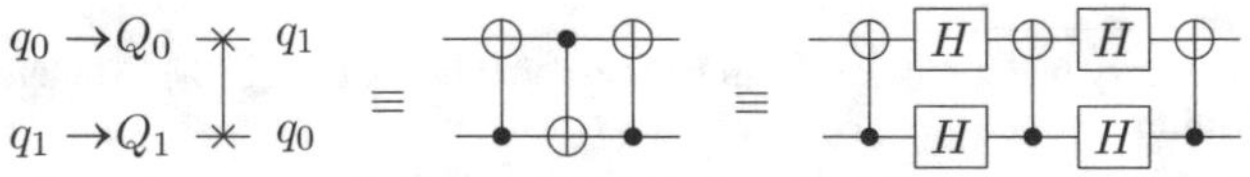

Fig. 15.1 Decomposition of a SWAP operation

However, inserting additional gates in order to satisfy the coupling-constraints drastically increases the number of operations—a significant drawback, which affects the fidelity of the quantum circuit since each gate has a certain error rate. Since each SWAP operation is composed of 7 elementary gates (cf. Fig. 15.1), particularly their number shall be kept as small as possible. Accordingly, this raises the question of how to derive a proper mapping of logical qubits to physical qubits while, at the same time, minimizing the number of added SWAP and H operations—an $\mathcal{NP}$-complete problem as recently proven in [18, 144].

Example 15.3 *Consider the quantum circuit composed of 5 CNOT gates shown in Fig. 15.2a and assume that the logical qubits q_0, q_1, q_2, q_3, q_4, and q_5 are, respectively, mapped to the physical qubits Q_0, Q_1, Q_2, Q_3, Q_{14}, and Q_{15} of IBM QX3 shown in Fig. 2.2c on Page 17. The first gate can be directly applied, because the coupling-constraints are satisfied. For the second gate, the direction has to be changed because a CNOT with control qubit Q_0 and target Q_1 is valid, but not vice versa. This can be accomplished by inserting Hadamard gates as shown in Fig. 15.2b. For the third gate, the mapping has to change. To this end, SWAP operations $SWAP(Q_1, Q_2)$ and $SWAP(Q_2, Q_3)$ are inserted to move logical qubit q_1 towards logical qubit q_4 (see Fig. 15.2b). Afterwards, q_1 and q_4 are mapped to the physical qubits Q_3 and Q_{14}, respectively, which allows applying the desired CNOT gate. Following this procedure for the remaining qubits eventually results in the circuit shown in Fig. 15.2b. The mapped circuit is composed of 51 elementary operations and has a depth of 36 when using a naive algorithm—a significant overhead that motivates research on improved approaches.*

Determining proper mappings has similarities with recent work on nearest-neighbor optimization of quantum circuits proposed in [131, 136, 137, 175–177, 185].[3] In that work, SWAP gates have also been applied to move qubits together in order to satisfy a coupling-constraint. However, these works consider simpler and artificial architectures with one-dimensional or two-dimensional layouts where

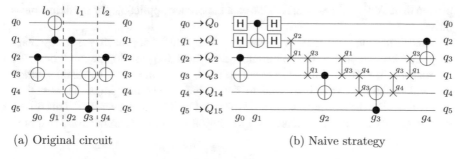

(a) Original circuit   (b) Naive strategy

Fig. 15.2 Mapping of a quantum circuit to IBM QX3

---

[3]These approaches utilize satisfiability solvers, search algorithms, or dedicated data structures to tackle the underlying complexity.

any two-qubit gate can be applied to adjacent qubits. The coupling-constraints to be satisfied for the IBM Q devices are much stricter with respect to what physical qubits may interact with each other and also what physical qubit may act as control and as target qubit. Furthermore, the parallel execution of gates (which is possible in IBM Q devices) is not considered by these approaches. Besides that, an approach has been proposed that utilizes temporal planning to compile quantum circuits to real architectures [161]. However, this approach is rather specialized to *Quantum Alternating Operator Ansatz* (QAOA [46]) circuits for solving the MaxCut problem and targets the architectures proposed by Rigetti (cf. [135]). Consequently, none of the approaches discussed above is directly applicable for the problem considered here.

In fact, there exist only very few algorithms that explicitly tackle the mapping problem for IBM Q devices [91, 144] and, thus, serve as alternative to IBM's own solution provided within its SDK Qiskit [33].[4] To encourage further development in this area, IBM even launched the *IBM Qiskit Developer Challenge* seeking for the best possible solution [80].

15.3   General Idea of the Presented Approaches

The following chapters present methods for mapping quantum circuits to the IBM Q devices that significantly enhance the current state of the art in different aspects. Chapter 16 (based on [170]) provides an exact approach (using a formal description of the mapping problem that is passed to a powerful reasoning engine) to solve the mapping problem by inserting the minimum number of additional H and SWAP operations. By this, a lower bound on the overhead is provided (when neglecting pre- and post-mapping optimizations), which is required to satisfy the coupling-constraints given by the quantum hardware—allowing to show that IBM's own solution often exceeds the minimal overhead by more than 100% (even for small instances). However, the exponential nature of the mapping problem (it has been proven to be $\mathcal{NP}$-complete) makes the exact approach applicable for small instances only.

This limitation—together with the fact that IBM's approach generates mapping that are far above the minimum—motivates the development of heuristic approaches. Therefore, Chap. 17 (based on [189, 190]) presents a heuristic solution that utilizes the A^* search method to determine proper mappings. This method allows reducing the overhead compared to Qiskit by 24.0% on average.[5] This difference in quality is mainly because IBM's solution randomly searches for a mapping

---

[4]Note that IBM's solution randomly searches (guided by heuristics) for mappings of the qubits at a certain point of time.

[5]Note that the approach has additionally been integrated into Qiskit to allow a fair comparison by utilizing the same post-mapping optimizations.

that satisfies the coupling-constraints—leading to a rather small exploration of the search space so that only rather poor solutions are usually found. In contrast, the presented heuristic approach aims for an optimized solution by exploring more suitable parts of the search space and additionally exploiting information of the circuit. More precisely, a look-ahead scheme is employed that considers gates that are applied in the near future and, thus, allows determining mappings which aim for a global optimum (instead of local optima) with respect to the number of SWAP operations.

Even though this heuristic approach allows outperforming Qiskit's mapping algorithm, it has some scalability issues when used for mapping certain random circuits for validating quantum computers [34], which also served as benchmarks in the *IBM Qiskit Developer Challenge* (a challenge for writing the best quantum-circuit compiler to encourage development). These circuits provide a worst-case scenario that heavily affects the efficiency of the presented heuristic approach. Therefore, a dedicated approach for this kind of circuits is presented in Chap. 18 (based on [199]), which explicitly considers their structure by using dedicated pre- and post-mapping optimizations. The resulting methodology has been declared as winner of the IBM Qiskit Developer Challenge, since it generated mapped/compiled circuits with at least 10% lower costs than the other submissions while generating them at least 6 times faster, and is currently being integrated into Qiskit by researchers from IBM. Besides that, all mapping approaches discussed in this book are publicly available at http://iic.jku.at/eda/research/ibm_qx_mapping, which also led to an official integration into Atos' QLM.

Chapter 16
Minimal and Close-to-Minimal Approaches

Abstract This chapter describes the first exact solution of the problem, utilizing powerful reasoning engines to determine a minimal solution. While this approach is obviously limited with respect to the size of the considered circuit and the underlying device (due to the exponential nature of the problem), it shows that existing solutions exceed the determined lower bound significantly—and that already for rather small circuits to be mapped.

Keywords Quantum circuits · Quantum gates · Mapping · Minimal cost · SWAP gates · IBM Q · Coupling map · Qubits · SAT solvers

Since the mapping problem has been proven to be $\mathcal{NP}$-complete, approaches that guarantee minimal or close-to-minimal results cannot scale for circuits of reasonable size—demanding for heuristic solutions as, e.g., provided by IBM within its SDK Qiskit. However, such heuristic solutions lead to a significant uncertainty about the quality of the achieved mappings, and whether it is worth to seek for improvement. Because of that, *exact* methods that can generate minimal (or at least close-to-minimal) results are essential for a substantial evaluation of heuristic methods—even if the optimum can only be determined for small instances.

This chapter (based on [170]) presented such solutions. To this end, *all* possible applications of the SWAP and H operations are considered that may influence the realization of an originally given circuit on an IBM Q device—a computationally very expensive task. In order to cope with this complexity, powerful reasoning engines such as solvers for Boolean satisfiability that can cope with large search spaces are used. Besides that, additional performance optimizations are considered that may lead to solutions which are not guaranteed to be minimal anymore, but can significantly reduce the solving time while still generating at least close-to-minimal solutions. This is confirmed by a conducted evaluation, which additionally shows that IBM's heuristic mapping solution exceeds the lower bound by more than 100% on average.

© Springer Nature Switzerland AG 2020
A. Zulehner, R. Wille, *Introducing Design Automation for Quantum Computing*,
https://doi.org/10.1007/978-3-030-41753-6_16

In the following, the exact approach is presented. To this end, Sect. 16.1 presents a symbolic formulation of the mapping problem that can be passed to reasoning engines to determine a mapped circuit with minimal overhead. To improve performance (at the cost of guaranteeing only close-to-minimal solutions) Sect. 16.2 presents techniques to reduce the considered search space. Eventually, Sect. 16.3 evaluates the presented approaches and compares them to IBM's own solution provided in Qiskit.

16.1 Determining a Minimal Solution

This section describes how to determine a minimal solution for the mapping problem. To this end, a symbolic formulation is introduced that describes the problem in terms of a Boolean function, which is eventually used to solve the problem by applying powerful and efficient reasoning engines (cf. Sect. 3.3).

The symbolic formulation of the considered problem—mapping quantum circuits to IBM Q devices using SWAP and H operations—requires defining variables which describe all possible applications of the SWAP and H operations. These operations affect how logical qubits from an originally given quantum circuit are mapped to the physical qubits of an IBM Q device. The mapping might be changed before each gate. Since only CNOT gates may cause violations of coupling-constraints, one-qubit gates are ignored when formulating the mapping problem.[1] This leads to the following symbolic formulation:

Definition 16.1 *Let* $G = g_0 g_1 \cdots g_k \cdots g_{|G|-1}$ *be a quantum circuit composed of* $|G|$ *CNOT gates. Each gate* $g_k = CNOT(q_c, q_t)$ *consists of a (logical) control qubit* q_c *and a (logical) target qubit* q_t *on which it operates. Furthermore, let* $q = \{q_0, \ldots, q_j, \ldots, q_{n-1}\}$ *be a set of n logical qubits that shall be mapped to the* $m \geq n$ *physical qubits in* $Q = \{Q_0, \ldots, Q_i, \ldots, Q_{m-1}\}$. *Finally, let* $CM \subseteq Q \times Q$ *be the description of the coupling map indicating what physical qubits can interact with each other on the given architecture and how (cf. Definition 2.3 on Page 16). Then, mapping variables* x_{ij}^k *with* $k \in \{0, \ldots, |G| - 1\}$, $i \in \{0, \ldots, m - 1\}$, *and* $j \in \{0, \ldots, n - 1\}$ *are introduced representing whether, before gate* $g_k \in G$, *the logical qubit* q_j *is mapped to the physical qubit* Q_i ($x_{ij}^k = 1$) *or not* ($x_{ij}^k = 0$).

Example 16.1 *Consider the circuit shown in Fig. 16.1a (and assume the one-qubit gates have been removed as shown in Fig. 16.1b). Then, Fig. 16.2 sketches a symbolic formulation for mapping the circuit to IBM QX4 (cf. Fig. 2.2b on Page 17), which represents all possible mappings of logical qubits to physical qubits. For example, the leftmost part of Fig. 16.2 represents the initial mapping of the logical qubits to the physical ones. Here, e.g., setting* $x_{02}^0 = 1$ *represents that logical qubit* q_2 *is mapped to physical qubit* Q_0 *right before gate* g_0.

---

[1]Note that this additionally reduces the overall complexity of the problem to be solved.

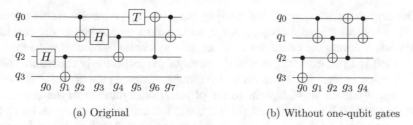

(a) Original   (b) Without one-qubit gates

Fig. 16.1 Quantum circuit to be mapped

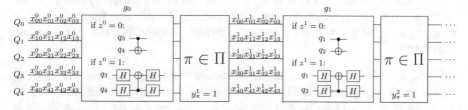

Fig. 16.2 Symbolic formulation for mapping the circuit shown in Fig. 16.1a

Passing this symbolic formulation to a reasoning engine would yield arbitrary assignments that most likely encode impossible/useless mappings (e.g., mapping several logical qubits to the same physical one). Hence, the assignment of variables has to be restricted so that only valid solutions are obtained. To this end, it has to be ensured that:

1. A bijective mapping between logical and physical qubits is conducted (i.e., each logical qubit is uniquely assigned to *exactly one* physical qubit and vice versa). This is ensured by

$$
\bigwedge_{k=0}^{|G|-1} \left(\bigwedge_{j=0}^{n-1} \left(\sum_{i=0}^{m-1} x_{ij}^{k} = 1 \right) \wedge \bigwedge_{i=0}^{m-1} \left(\sum_{j=0}^{n-1} x_{ij}^{k} \leq 1 \right) \right). \tag{16.1}
$$

2. All gates only act on physical qubits that have a corresponding entry in the coupling map of the considered architecture or an entry where control and target qubit are switched.[2] This is ensured by

$$
\bigwedge_{g_k=CNOT(q_c,q_t)\in G} \left(\bigvee_{(Q_i,Q_j)\in CM} \left(x_{ic}^{k} \wedge x_{jt}^{k} \right) \vee \left(x_{it}^{k} \wedge x_{jc}^{k} \right) \right). \tag{16.2}
$$

---

[2]Note that also entries where the control and target qubit are switched are considered since this can be handled by inserting H gates. This way, the reasoning engines get to decide which sequence of SWAP and H gates yield the cheapest global mapping.

Adding these restrictions and passing the resulting symbolic formulation to a reasoning engine eventually yields a valid solution. Moreover, the resulting formulation covers the entire search space in a symbolic fashion. Having this, all that is left is a proper description of the costs of the respectively chosen mapping. As reviewed in Sect. 15.2, these costs accumulate from (1) the costs for changing the mapping of the logical qubits to the physical ones throughout the circuit (by inserting SWAP operations) and (2) the costs for switching the control and the target qubit for CNOT gates (by inserting H operations). The former one (changes on the mapping) may be applied before each CNOT gate (except the first one, which defines the initial mapping and, hence, can be set arbitrarily anyway); the latter one (switching the control/target qubits) may be required for each gate. To properly describe this within the symbolic formulation, the following variables are introduced:

Definition 16.2 *Let* $0 \leq k < |G|$ *be the index of gates* g_k *in a quantum circuit, m the number of physical qubits in the considered quantum device, and* $\pi \in \Pi$ *a permutation of m elements that indicates how the state of the physical qubits is permuted (eventually realized by inserting SWAP operations). Then, the* permutation variables y_π^k *indicate whether the permutation* π *is applied before gate* g_k *(*$y_\pi^k = 1$*) or not (*$y_\pi^k = 0$*). Furthermore, the* switching variables z^k *indicate whether the direction of the CNOT gate* g_k *is switched (*$z^k = 1$*) or not (*$z^k = 0$*).*

Example 16.2 *Consider again symbolic the formulation shown in Fig. 16.2. The spots in the circuit where the mapping may change are sketched by boxes labeled* π*. Here, the variable assignment before and after (i.e., the assignments of* x_{ij}^{k-1} *and* x_{ij}^k*) may change according to a permutation* $\pi \in \Pi$ *(eventually to be represented by* y_π^k*). Furthermore, in each gate* g_k*, the* z^k*-variables define whether the direction of the CNOT gate is switched or not.*

These variables allow describing what permutation $\pi \in \Pi$ is applied to the states of the physical qubits of the circuit lines before each gate $g_k \in G$ by introducing

$$\bigwedge_{k=1}^{|G|-1} \left(\bigwedge_{\pi \in \Pi} \left(\bigwedge_{i=0}^{m-1} \bigwedge_{j=0}^{n-1} \left(x_{ij}^{k-1} = x_{\pi(i)j}^k \right) \right) \Leftrightarrow y_\pi^k \right). \tag{16.3}$$

In fact, this ensures that y_π^k is set to 1 iff the assignment of the variables x_{ij}^{k-1} and x_{ij}^k indeed describe a change of the mapping defined by π.[3]

Similarly, it is described for each CNOT gate g_k whether the direction of the control and target qubit is switched by introducing

---

[3]If $n < m - 1$, π cannot be determined uniquely. Then, a left-handed implication is required instead of an equivalence in Eq. (16.3)—in conjunction with a constraint that only one variable y_π^k is assigned to 1. For sake of clarity of Eq. (16.3), $n = m$ is assumed.

$$\bigwedge_{g_k=CNOT(q_c,q_t)\in G} \left(\bigvee_{(Q_i,Q_j)\in CM} \left(x_{it}^k \wedge x_{jc}^k \right) \right) \Leftrightarrow z^k. \qquad (16.4)$$

In fact, this ensures that z^k is set to 1 iff, for gate g_k, the control qubit is set to position j and the target qubit is set to position i although, according to the coupling map, it has to be vice versa (i.e., control and target qubits are switched).

Satisfying all of the constraints introduced above yields a valid mapping of the originally given circuit to the desired architecture while, at the same time, the costs are determined by

$$\mathcal{F} = \sum_{k=1}^{|G|-1} \sum_{\pi \in \Pi} \left(7 \cdot \text{swaps}(\pi) \, y_\pi^k \right) + \sum_{k=0}^{|G|-1} \left(4 \cdot z^k \right). \qquad (16.5)$$

Here, $\text{swaps}(\pi)$ defines the number of SWAP operations needed to realize the permutation π. This has to be determined for each permutation $\pi \in \Pi$—a process, which needs to be conducted only once and can be done, e.g., by using an exhaustive search for the considered architectures. By this, whenever the reasoning engine chooses a mapping which eventually creates a permutation π before gate g_k, Eq. (16.3) sets $y_\pi^k = 1$ and, hence, adds the corresponding costs (7 gates for each SWAP operation; cf. Sect. 15.2) to the overall costs $\mathcal{F}$. Similarly, whenever the reasoning engine chooses a mapping which switches the direction of a CNOT gate g_k, Eq. (16.4) sets $z^k = 1$ and, hence, adds the corresponding costs (4 H operations; cf. Sect. 15.2) to the overall costs $\mathcal{F}$.

Passing the eventually resulting symbolic formulation to a reasoning engine allows determining a valid mapping together with the associated costs (i.e., the number of additionally required elementary operations). Since in the minimum costs are of interest, the cost function $\mathcal{F}$ needs to be further restricted. One direct solution could be, e.g., simply setting $\mathcal{F}$ to a fixed value and, e.g., by applying a binary search approach towards the minimum. However, since many reasoning engines additionally allow the consideration of an objective function (cf. Definition 3.1 on Page 25), the most efficient way is to simply add the objective $\min : \mathcal{F}$ to the resulting instance—enforcing the reasoning engine not only to determine a satisfying assignment (representing a valid mapping) but, at the same time, also to minimize $\mathcal{F}$.

Example 16.3 *Passing the symbolic formulation sketched in Fig. 16.2 together with all constraints and the objective function to a reasoning engine eventually yields a mapping (and a corresponding addition of SWAP and H operations) as shown in Fig. 16.3. This circuit provides a realization of the originally given circuit from Fig. 16.1a which is applicable for the IBM Q device specified by the coupling map shown in Fig. 2.2b on Page 17 and, at the same time, yields minimum costs caused by additionally required SWAP and H operations ($\mathcal{F} = 4$).*

Fig. 16.3 Resulting circuit (with minimal SWAP/H costs)

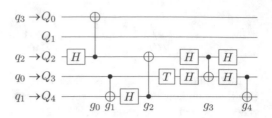

16.2 Improving Performance

Determining minimal solutions obviously is the desired way to go. However, even powerful reasoning engines cannot circumvent the $\mathcal{NP}$-complete complexity of the problem. In this regard, the methodology presented in the previous section allows for several performance improvements. In fact, the reasoning engine only needs to determine a "minimal" assignment for the x_{ij}^k-variables (the variables y_π^k and z^k can be ignored, since they are only used for formulating the costs and their assignments can directly be deduced from the x_{ij}^k-variables). For a quantum circuit composed of n logical qubits and $|G|$ CNOT gates to a device with m physical qubits, this leads to a total of $n \cdot m \cdot |G|$ variables to be assigned and, hence, an overall search space of $2^{n \cdot m \cdot |G|}$ which can be easily restricted by adding further constraints to the x_{ij}^k-variables. While this may lead to solutions which are not guaranteed to be minimal anymore, it can significantly speed up the solving time while, at the same time, remaining close-to-minimal (as also confirmed by an evaluation summarized in Sect. 16.3). This section shows possible improvements in this regard.

16.2.1 Considering Subsets of Physical Qubits

A scenario frequently occurs where the number n of logical qubits of a given quantum circuit to be mapped is smaller than the number m of physical qubits provided by the architecture (i.e., where $n < m$). Then, obviously, not all physical qubits are required. In fact, this allows considering only a subset of n physical qubits while ignoring the remaining $m - n$ ones. Since the number of physical qubits to consider contributes to the search space in an exponential fashion, restricting this number yields substantial simplifications. In order to remain as close as possible to the minimal solution, one can try out all $\binom{m}{n}$ possible subsets of qubits to consider and solve the respectively resulting (smaller) instances separately. This reduces the overall search space to $\binom{m}{n} 2^{n^2 \cdot |G|}$.

Example 16.4 *Consider again the symbolic instance for mapping a four-qubit quantum circuit to a five qubit architecture as sketched in Fig. 16.2. By considering only four physical qubits in the mapping procedure, the overall search space for a single instance reduces from $2^{4 \cdot 5 \cdot 5} = 2^{100}$ to $2^{4^2 \cdot 5} = 2^{80}$. Even if all $\binom{5}{4} = 5$*

possible subsets of physical qubits are considered separately, this still yields a significant reduction of the overall search space.

The search space can be further reduced by checking whether some of the physical qubits in a subset are isolated from others (this can be done in $O(n)$ time). If so, the instance for this subset does not have to be passed to the reasoning engine as no solution can be found anyway.

Example 16.5 *Assume that the circuit over four qubits considered thus far shall be mapped to IBM QX4 shown in Fig. 2.2b on Page 17. Then, all subsets of physical qubits that have to be checked should contain Q_2, since no connected sub-graph composed of four nodes without Q_2 is possible. This reduces the number of instances that are passed to the reasoning engine from $\binom{5}{4} = 5$ to $\binom{4}{3} = 4$.*

16.2.2 Restricting the Possible Permutations

Thus far, permutations $\pi \in \Pi$ of the mapping are allowed before each gate (except the first one where an arbitrary initialization can be chosen). While this guarantees minimality (since all possible solutions are considered), this substantially contributes to the complexity. In many cases, however, valid and cheap mappings are still possible if permutations of mappings are allowed not before *all* gates $g \in G$, but only before a subset $G' \subseteq G \setminus \{g_0\}$ of them. With $|G'|$ being significantly smaller than G, this reduces the overall search space to $2^{n \cdot m \cdot (|G'|+1)}$. While, applying this idea, G' can be chosen arbitrarily, a smaller G' leads to a larger performance improvement but also a more restricted instance (yielding solutions that might be far from minimal or even instances for which no valid mapping can be determined anymore). In this section, the following strategies for defining G' are considered:

- *Disjoint qubits*, which exploits the fact that gates acting on disjoint sets of qubits can always be mapped in a way that no intermediate permutations are required. To this end, the quantum circuit is clustered into sequences of gates acting on disjoint sets of qubits and permutations are only allowed before each of those sequences.
- *Even gates*, which allows permutations only before gates with an even index (except for g_0). Here, it is still guaranteed that a valid mapping can be determined since either (1) the gates operate on the disjoint sets of qubits as discussed above, (2) the gates share both qubits, or (3) the gates share one qubit (and there exists at least one qubit that can interact with two other qubits).
- *Qubit triangle*, which exploits the structure of architectures whose coupling map forms "triangles" of physical qubits as in case of, e.g., Q_0, Q_1, and Q_2 in Fig. 2.2b on Page 17. Here, the circuit can be clustered into sequences of gates where each sequence acts on at most three qubits. Then, each such sequence of gates is mapped to a triangle as described above and permutations are only required before each of those sequences.

Example 16.6 *Consider the quantum circuit shown in Fig. 16.1b. Applying the strategies discussed above yields the following subsets G':*

- Disjoint qubits: $G' = \{g_2, g_3, g_4\}$, *since the gates g_0 and g_1 operate on disjoint qubits (saving the permutation between g_0 and g_1).*
- Even gates: $G' = \{g_2, g_4\}$, *since they constitute the even gates in the circuit.*
- Qubit triangle: $G' = \{g_1\}$, *since all gates g_1, g_2, g_3, and g_4 operate on only three qubits and, hence, can be mapped to one of the "triangles" of the architecture without the need for further permutations. That is, only a permutation prior to g_1 needs to be considered.*

As can be seen, these strategies yield much more restrictive applications of permutations. While this substantially increases the performance of the solving process, it does not harm minimality in this case (but may for other circuits as evaluated in the next section).

16.3   Evaluation

The presented methods for mapping quantum circuits to IBM Q devices using the minimal or close-to-minimal number of SWAP and H operations have been implemented in C++. As reasoning engine, the *Z3* solver [39] has been utilized. Afterwards, an extensive evaluation was conducted using quantum circuits taken from [35, 173] to be mapped to IBM QX4 [77]. The evaluation has been conducted on an Intel Core i7 machine with 4 GHz and 64 GB of main memory.

The left part of Table 16.1 thereby provides the results obtained by the first part of the evaluation—aiming for evaluating the effect of the performance improvements discussed in Sect. 16.2. More precisely, the first three columns describe the name of the considered quantum circuit, the number of logical qubits n, and the *original costs* of the circuit (i.e., the number of one-qubit gates plus the number of CNOT gates before mapping). The remaining columns list the costs c (i.e., the number of gates) of the obtained circuit and the run-time t (in CPU seconds) required by Z3 when applying the method guaranteeing minimality discussed in Sect. 16.1 as well as the adapted methods additionally incorporating the performance improvements discussed in Sect. 16.2. For the adapted versions, the difference to the minimum (i.e., Δ_{min}) is listed in parenthesis. Moreover, for each strategy that limits the number of permutations, the column denoted $|G'|$ indicates how many permutations are allowed.

As can be seen by these results, determining a mapping with the minimum number of SWAP and H gates is quite expensive—only solutions for instances with rather few CNOT gates can be determined (which is not surprising since the underlying problem is $\mathcal{NP}$-complete). When considering only a subset of the physical qubits (cf. Sect. 16.2.1), a significant reduction of the run-time for benchmarks with 3 or 4 qubits is observed, while still preserving minimality.

Table 16.1 Evaluation

| Benchmark | n | Original costs | Min. (Sect. 16.1) | | Perf. Opt. (Sect. 16.2.1) | | Performance optimized (Sect. 16.2.2) | | | | | | | | | IBM [33] |
| | | | | | | | Disjoint qubits | | | Even gates | | | Qubit triangle | | | |
			c_{min}	$t\,[s]$	$c\,(\Delta_{min})$	$t\,[s]$	$|G'|$	$c\,(\Delta_{min})$	$t\,[s]$	$|G'|$	$c\,(\Delta_{min})$	$t\,[s]$	$|G'|$	$c\,(\Delta_{min})$	$t\,[s]$	$c\,(\Delta_{min})$
3_17_13	3	19 + 17 = 36	59	29	59(+0)	0	17	59(+0)	0	9	60(+1)	0	1	60(+1)	0	80(+21)
ex-1_166	3	10 + 9 = 19	31	5	31(+0)	0	9	31(+0)	0	5	31(+0)	0	1	31(+0)	0	39(+8)
ham3_102	3	9 + 11 = 20	36	10	36(+0)	0	11	36(+0)	0	6	36(+0)	0	1	36(+0)	0	48(+12)
miller_11	3	27 + 23 = 50	82	231	82(+0)	0	23	82(+0)	0	12	82(+0)	0	1	82(+0)	0	82(+0)
4gt11_84	4	9 + 9 = 18	34	7	34(+0)	0	9	34(+0)	0	5	34(+0)	0	2	34(+0)	0	37(+3)
rd32-v0_66	4	18 + 16 = 34	63	281	63(+0)	35	16	63(+0)	35	8	63(+0)	1	2	72(+9)	0	101(+38)
rd32-v1_68	4	20 + 16 = 36	65	276	65(+0)	35	16	65(+0)	36	8	65(+0)	1	2	74(+9)	0	99(+34)
4gt11_82	5	9 + 18 = 27	62	133	62(+0)	137	18	62(+0)	139	9	62(+0)	3	5	62(+0)	1	77(+15)
4gt11_83	5	9 + 14 = 23	49	17	49(+0)	17	14	49(+0)	18	7	50(+1)	1	3	50(+1)	0	65(+16)
4gt13_92	5	36 + 30 = 66	109	528	109(+0)	533	29	109(+0)	199	15	110(+1)	10	9	110(+1)	5	126(+17)
4mod5-v0_19	5	19 + 16 = 35	64	256	64(+0)	264	16	64(+0)	255	8	68(+4)	2	3	69(+5)	0	109(+45)
4mod5-v0_20	5	10 + 10 = 20	35	10	35(+0)	10	10	35(+0)	11	5	35(+0)	0	3	35(+0)	0	64(+29)
4mod5-v1_22	5	10 + 11 = 21	40	7	40(+0)	7	10	40(+0)	9	6	40(+0)	0	3	43(+3)	0	52(+12)
4mod5-v1_24	5	20 + 16 = 36	63	54	63(+0)	55	16	63(+0)	56	8	63(+0)	3	3	63(+0)	0	98(+35)
alu-v0_27	5	19 + 17 = 36	63	74	63(+0)	73	16	63(+0)	38	9	63(+0)	2	3	67(+4)	0	101(+38)

(continued)

Table 16.1 (continued)

Benchmark	n	Original costs	Min. (Sect. 16.1)		Perf. Opt. (Sect. 16.2.1)		Performance optimized (Sect. 16.2.2)									IBM [33]
							Disjoint qubits			Even gates			Qubit triangle			
			c_{min}	$t\,[s]$	$c\,(\Delta_{min})$	$t\,[s]$	$|G'|$	$c\,(\Delta_{min})$	$t\,[s]$	$|G'|$	$c\,(\Delta_{min})$	$t\,[s]$	$|G'|$	$c\,(\Delta_{min})$	$t\,[s]$	$c\,(\Delta_{min})$
alu-v1_28	5	19 + 18 = 37	64	94	64(+0)	92	17	64(+0)	44	9	67(+3)	10	3	68(+4)	0	123(+59)
alu-v1_29	5	20 + 17 = 37	64	351	64(+0)	355	16	64(+0)	119	9	64(+0)	3	3	68(+4)	0	104(+40)
alu-v2_33	5	20 + 17 = 37	64	42	64(+0)	44	17	64(+0)	44	9	64(+0)	4	4	64(+0)	0	99(+35)
alu-v3_34	5	28 + 24 = 52	90	719	90(+0)	727	24	90(+0)	724	12	91(+1)	10	4	91(+1)	0	178(+88)
alu-v3_35	5	19 + 18 = 37	64	103	64(+0)	101	17	64(+0)	74	9	64(+0)	3	3	68(+4)	0	121(+57)
alu-v4_37	5	19 + 18 = 37	64	119	64(+0)	121	17	64(+0)	43	9	64(+0)	6	3	68(+4)	0	110(+46)
mod5d1_63	5	9 + 13 = 22	48	14	48(+0)	13	11	48(+0)	8	7	48(+0)	5	5	48(+0)	1	98(+50)
mod5mils_65	5	19 + 16 = 35	64	96	64(+0)	98	16	64(+0)	94	8	65(+1)	1	3	65(+1)	0	108(+44)
qe_qft_4	5	44 + 27 = 71	94	136	94(+0)	135	19	94(+0)	21	14	94(+0)	9	16	94(+0)	12	115(+21)
qe_qft_5	5	69 + 38 = 107	135	401	135(+0)	395	26	135(+0)	21	19	139(+4)	107	24	145(+10)	48	163(+28)

n: number of logical qubits

Original costs: number of one-qubit gates plus number of CNOT gates before mapping

c: costs (number of operations) of the mapped circuit

Δ_{min}: difference to minimum costs

t: run-time in seconds

Limiting the number of permutation has a tremendous effect on the run-time. In fact, the run-time required to solve an instance directly correlates with $|G'|$. However, limiting the number of permutations too much generates rather poor results regarding minimality. For the benchmarks considered in the evaluation, the strategy *disjoint qubits* always generates results with minimum cost, whereas the strategy *qubit triangle* yields the poorest results regarding minimality. Still, all strategies provide alternatives delivering solutions that are close-to-minimal within acceptable run-time.

In the second part of the evaluation, the obtained minimal and close-to-minimal solutions are compared to the mapping algorithm provided by IBM's SDK Qiskit [33]. This allows evaluating how far existing heuristic approaches deviate from the optimum. To this end, the last column of Table 16.1 lists the number of gates in the obtained circuit when utilizing the mapper available in Qiskit 0.4.15 (the run-time is not listed since all mappings were determined within a second).[4] Since the mapping algorithm in Qiskit is probabilistic, the observed minimum of 5 runs is listed.

As can be seen, the heuristic approach utilized in Qiskit can be improved significantly. Considering the benchmarks *alu-v3_35* and *mod5d1_63*, IBM's algorithm is 89.1 and 104.2% above the practical minimum that can be reached, respectively. On average, IBM's solution yields circuits that are 45% above the minimum (by means of gate count). Considering only the number of gates added during the mapping (and not the complete mapped circuit), Qiskit's solutions are 109.2% above the minimum given by $\mathcal{F}$ on average—doubling the overhead required for mapping a circuit.

Hence, even though the exact approach presented in this chapter is only applicable for mapping small quantum circuits, it shows that there is a lot of room for improvement of heuristic approaches—further motivating research on this topic.

---

[4]Note that, to ensure a fair comparison, only the actual mapping process of Qiskit is considered—neglecting the decomposition as well as pre- or post-mapping optimizations.

Chapter 17
Heuristic Approach

Abstract This chapter presents a heuristic approach for the considered mapping problem. To this end, the A^* search algorithm is utilized. The proposed method is one of the first approaches addressing the practically relevant research question of how to efficiently map quantum circuits to IBM Q devices (or NISQ devices in general).

Keywords Quantum circuits · Quantum gates · Mapping · A^* search · Layers · SWAP gates · IBM Q · Coupling map · Qubits

The exact method described in the previous chapter unveils significant optimization potential for IBM's own solution for mapping quantum circuits to the IBM Q devices. Together with the infeasibility of exact approaches due to the exponential complexity of the underlying problem (cf. Chap. 16), this motivates research on a new heuristic approach that utilizes knowledge gained in design automation from the conventional domain.

This chapter (based on [189, 190]) introduces a multi-step approach, which utilizes a depth-based partitioning and A^* as underlying search algorithm to solve the mapping problem. Moreover, utilizing optimizations such as a look-ahead scheme and the ability to determine the initial mapping of the qubits throughout the mapping process (instead of fixing the initial mapping at the beginning of the algorithm) leads to significant reductions in run-time as well as in the number of gates of the mapped circuit compared to IBM's approach. The resulting methodology is generic, i.e., it can be directly applied to all existing IBM Q devices as well as to similar architectures, which may come up in the future. Finally, the presented heuristic approach has been integrated into IBM's SDK *Qiskit* [33]—allowing for a more realistic performance evaluation since post-mapping optimizations provided by IBM are additionally considered.

A. Zulehner, R. Wille, *Introducing Design Automation for Quantum Computing*,
https://doi.org/10.1007/978-3-030-41753-6_17

In the following, the heuristic approach is described. To this end, Sect. 17.1 describes how to efficiently satisfy the coupling-constraints utilizing A^* search. Afterwards, Sect. 17.2 presents optimizations to further reduce the costs of the mapped circuit, while Sect. 17.3 evaluates the heuristic approach and compares it to IBM's solution.

17.1 Efficiently Satisfying Coupling-Constraints

This section proposes an efficient method for mapping a given quantum circuit (which has already been decomposed into a sequence of elementary gates as described in Sect. 15.1) to IBM Q devices. The main objective is to minimize the number of elementary gates, which are added in order to make the mapping coupling-constraint-compliant. Two main steps are employed: First, the given circuit is partitioned into layers that can be realized in a coupling-constraint-compliant fashion. Afterwards, for each of these layers, a particular compliant mapping is determined which requires as few additional gates as possible. In the following subsections, both steps are described in detail.

17.1.1 Partitioning the Circuit into Layers

As mentioned in Chap. 15, the mapping from logical qubits to physical ones may change over time in order to satisfy all coupling-constraints, i.e., the mapping may have to change before a CNOT can be applied. Since each change of the mapping requires additional SWAP operations, the presented approach aims for conducting these changes as rarely as possible. To this end, gates that can be applied concurrently are combined into the so-called *layers* (i.e., sets of gates). A layer l_i contains only gates that act on distinct sets of qubits. Furthermore, this allows determining a mapping such that the coupling-constraints for all gates $g_j \in l_i$ are satisfied at the same time. The layers are formed in a greedy fashion, i.e., a gate is added to the layer l_i where i is as small as possible. In the circuit diagram representation, this means to move all gates to the left as far as possible without changing the order of gates that share a common qubit. Note that the depth of a circuit is equal to the number of layers of a circuit.

Example 17.1 *Consider again the quantum circuit shown in Fig. 15.2a on Page 163. The gates of the circuit can be partitioned into three layers $l_0 = \{g_0, g_1\}$, $l_1 = \{g_2, g_3\}$, and $l_2 = \{g_4\}$ (indicated by the dashed lines in Fig. 15.2a on Page 163).*

To satisfy all coupling-constraints, the logical qubits of each layer l_i have to be mapped to physical ones. Since the resulting mapping for layer l_i is not necessarily equal to the mapping determined for the previous layer l_{i-1}, SWAP operations need to be inserted that permute the logical qubits from the mapping for layer l_{i-1} to the desired mapping for layer l_i. In the following, this sequence of SWAP operations is denoted *permutation layer* π_i. The mapped circuit is then an interleaved sequence of the layers l_i of the original circuit, and the corresponding permutation layers π_i, i.e., $l_0\pi_1 l_1\pi_2 l_2 \ldots$.

17.1.2 Determining Compliant Mappings for Layers

Let $\sigma_j^i : \{q_0, q_1, \ldots q_{n-1}\} \to \{Q_0, Q_1, \ldots Q_{m-1}\}$ describe mappings from logical qubits of the quantum circuit to the physical ones of the target device for layer l_i. Then, given an initial mapping σ_0^i, a mapping $\hat{\sigma}_0^i$ shall be found that satisfies the coupling-constraints for all gates $g \in l_i$ while being established from σ_0^i with minimum costs, i.e., with the minimum number of additionally required elementary operations. The mapping σ_0^i is obtained from the solution found for the previous layer l_{i-1}, i.e., $\sigma_0^i = \hat{\sigma}^{i-1}$ (for l_0, a randomly generated initial mapping that satisfies all coupling-constraints for the gates $g \in l_0$ is used). In the worst case, determining $\hat{\sigma}^i$ requires the consideration of $m!/(m - n)!$ possibilities (where m and n are the number of physical qubits and logical qubits, respectively)—an exponential complexity. The presented approach aims to cope with this complexity by applying an A^* search algorithm (cf. Sect. 3.2).

To use the A^* algorithm for the considered search problem, an expansion strategy for a node (i.e., a mapping σ_j^i) as well as an admissible heuristic function $h(x)$ to estimate the distance of a node to a goal node (i.e., the mapping $\hat{\sigma}^i$) are required. Given a mapping σ_j^i, all possible successor mappings σ_h^i are determined by employing all possible combinations of SWAP gates that can be applied concurrently.[1] The fixed costs of all these successor states σ_h^i is then $f(\sigma_h^i) = f(\sigma_j^i) + 7 \cdot \#SWAPS$ since each SWAP gate is composed of 7 elementary operations (3 CNOTs and 4 Hadamard operations). Note that the expansion strategy is restricted to SWAP operations that affect at least one qubit that occurs in a CNOT gate $g \in l_i$ on layer l_i. This is justified by the fact that only these qubits influence whether or not the resulting successor mapping is coupling-constraint-compliant.

Example 17.2 *Consider again the quantum circuit shown in Fig. 15.2a on Page 163 and assume a mapping to IBM QX3 (cf. Fig. 2.2c on Page 17) is desired for layer $l_1 = \{g_2, g_3\}$. In the previous layer l_0, the logical qubits q_1, q_3, q_4, and q_5 have been mapped to the physical qubits Q_0, Q_3, Q_{14}, and Q_{15}, respectively (i.e., $\hat{\sigma}^0$). This initial mapping $\sigma_0^1 = \hat{\sigma}^0$ does not satisfy the coupling-constraints for the gates in l_1. Since the CNOTs of l_1 only operate on four qubits, σ_0^i has 51 successors σ_j^i.*

---

[1] Note that multiple SWAP gates are applied concurrently in order to minimize the circuit depth as second criterion (if two solutions require the same number of additional operations).

As mentioned in Sect. 3.2, a heuristic function that does not overestimate the real costs (i.e., the minimum number of additionally inserted elementary operations) is required to obtain an optimal mapping (i.e., the mapping with the fewest additionally required elementary operations that satisfies all coupling-constraints). The real minimum costs for an individual CNOT gate $g \in l_i$ can be easily determined given σ_j^i. Assume that the control and target qubit of g are mapped to the physical qubits Q_c and Q_t (which is given by σ_j^i). Then the shortest path $\hat{p}$ from Q_c to Q_t (following the arrows in the coupling map[2]) determines the costs of a CNOT gate. More precisely, the costs of the CNOT gate $h(g, \sigma_j^i) = (|\hat{p}|-1) \cdot 7$ are determined by the length of this shortest path $|\hat{p}|$. In fact, $(|\hat{p}| - 1)$ SWAP operations are required to move the control and target qubits of g towards each other. If none of the arrows of the path $\hat{p}$ on the coupling map (representing that a CNOT can be applied) points into the desired direction, the true minimum costs increase by 4, since 2 Hadamard operations are required before and after the CNOT to change its direction.

The heuristic costs of a mapping σ_j^i are determined by the real costs of each CNOT gate $g \in l_i$ in layer l_i. Simply summing them up might overestimate the true cost, because one SWAP operation might reduce the distance of the control and target qubits for more than one CNOT of layer l_i. Since this might lead to a suboptimal solution $\hat{\sigma}^i$, the heuristic costs of a state σ_j^i are determined by $h(\sigma_j^i) = \max_{g \in l_i} h(g, \sigma_j^i)$, i.e., the maximum of the true costs of the CNOTs in layer l_i.

Example 17.2 (Continued) *The logical qubits q_1 and q_4 are mapped to the physical qubits $\sigma_0^1(q_1) = Q_1$ and $\sigma_0^1(q_4) = Q_{14}$, respectively. Since the shortest path on the coupling map is $\hat{p} = Q_1 \rightarrow Q_2 \rightarrow Q_3 \rightarrow Q_{14}$ (cf. Fig. 2.2c on Page 17), the true minimum costs for g_2 is $h(g_2, \sigma_0^1) = 2 \cdot 7 = 14$. Analogously, the costs of g_3 are $h(g_3, \sigma_0^1) = 7$—resulting in overall heuristic costs of $h(\sigma_0^1) = \max(14, 7) = 14$ for the initial mapping. Following the A^* algorithm outlined above, a mapping $\hat{\sigma}^1$ is determined which maps the logical qubits q_0, q_1, q_2, q_3, q_4, and q_5 to the physical qubits Q_0, Q_2, Q_1, Q_4, Q_3, and Q_5 by inserting two SWAP operations (as depicted in Fig. 17.1). Applying the algorithm also for mapping layer l_2, the circuit shown in Fig. 17.1 results. This circuit is composed of 37 elementary operations and has depth 15.*

17.2   Optimizations

A^* allows efficiently determining an optimal mapping (by means of additionally required operations) for each layer. However, the algorithm presented in Sect. 17.1.2 considers only a single layer when determining $\hat{\sigma}^i$ for layer l_i.

---

[2]The direction of the arrow does not matter since a SWAP can be applied between two physical qubits iff a CNOT can be applied.

Fig. 17.1 Circuit resulting from locally optimal mappings

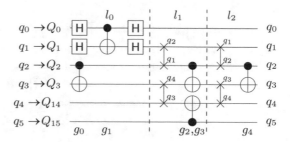

One way to optimize the heuristic solution is to employ a look-ahead scheme, which incorporates information from the following layers to the cost function. To this end, only the heuristics change, which to estimate the costs for reaching a mapping that satisfies all coupling-constraints from the current one. The approach described in Sect. 17.1.2 uses the maximum of the costs for each CNOT gate in layer l_i to estimate the true remaining cost. To employ a look-ahead scheme, an additional estimate for layer l_{i+1} is determined. The overall heuristic that guides the search algorithm towards a solution is then the sum of both estimates.

To incorporate the look-ahead scheme, the heuristics discussed in Sect. 17.1.2 change. Instead of taking the maximum of the CNOTs in the current layer, the costs of all CNOTs in two layers (the current layer and the look-ahead layer) are summed up, i.e., $h(\sigma_j^i) = \sum_{g \in l_i \cup l_{i+1}} h(g, \sigma_j^i)$. As discussed above, this might lead to an over-estimation of the true remaining costs for reaching a goal node. Thus, the solution is not guaranteed to be locally optimal. However, this is not desired anyways, since locally suboptimal solutions are desired in order to find cheaper mappings for the following layers—resulting in smaller mapped circuits.

Example 17.3 *Consider again the quantum circuit shown in Fig. 15.2a on Page 163 and assume that the logical qubits q_0, q_1, q_2, q_3, q_4, and q_5 are mapped to the physical qubits Q_0, Q_1, Q_2, Q_3, Q_{14}, and Q_{15}, respectively. Using the look-ahead scheme discussed above will not determine the locally optimal solution with costs of 14 for layer l_1 (as discussed in Example 17.2), but a mapping $\hat{\sigma}^1$ that satisfies all coupling-constraints with costs of 22 (as show in Fig. 17.2). The additional costs of 8 result since, after applying two SWAP gates (cf. Fig. 17.2), the directions of both CNOTs of layer l_1 have to change. However, this mapping also satisfies all coupling-constraints for layer l_2, which means that the remaining CNOT g_4 can be applied without adding further SWAPs. The resulting circuit is composed of a total of 31 elementary operations and has depth of 12 (as shown in Fig. 17.2; gates g_2 and g_3 can be applied concurrently). Consequently, the look-ahead scheme results in a cheaper mapping than the "pure" methodology presented in Sect. 17.1.2 and yielding the circuit shown in Fig. 17.1.[3]*

---

[3]Note that the graphical representation seems to be larger in Fig. 17.2. However, this is caused by the fact that the SWAP operations are not decomposed (cf. Fig. 15.1 on Page 162) in order to maintain readability.

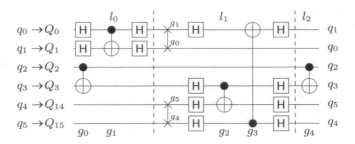

Fig. 17.2 Circuit generated when using the look-ahead scheme

Fig. 17.3 Resulting mapping

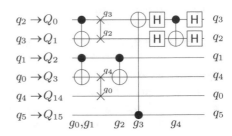

Besides the look-ahead scheme, the methodology is further improved by not starting with a random mapping for layer l_0. Instead, partial mappings σ_j^i are used— starting with an empty mapping σ_0^0 (i.e., none of the logical qubits is mapped to a physical one). Then, before starting to search a mapping for layer l_1, it is checked whether the qubits that occur in the CNOTs $g \in l_i$ have already been mapped for one of the former layers. If not, one of the "free" physical qubits (i.e., a physical qubit no logical qubit is mapped to) can be chosen arbitrarily. Obviously, the physical qubit is chosen so that the costs for finding $\hat{\sigma}^i$ are as small as possible.

This scheme provides the freedom to evolve the initial mapping throughout the mapping process, rather than starting with an initial mapping that might be non-beneficial with respect to the overall number of elementary operations.

Example 17.4 *Optimizing the methodology with a partial mapping that is initially empty results in the circuit shown in Fig. 17.3. This circuit is composed of 23 elementary operations and has depth 10 (gates g_2 and g_3 can be applied concurrently).*

17.3   Evaluation

Taking all considerations and methods discussed above into account led to the development of a mapping methodology which maps quantum circuits to an **IBM Q** device such that all coupling-constraints are satisfied. To allow for comparisons against IBM's SDK *Qiskit* [33], the mapping methodology presented in this chapter has been implemented in C++ and integrated into Qiskit 0.4.15. The adapted version of *Qiskit* as well as a stand-alone version of the methodology are publicly available at http://iic.jku.at/eda/research/ibm_qx_mapping.

This section compares the efficiency of the resulting scheme to the mapping approach implemented in *Qiskit* [33]. To this end, several functions taken from RevLib [173] as well as quantum algorithms written in Quipper [57] or the Scaffold language [4] (and pre-compiled by the ScaffoldCC compiler [82]) have been considered as benchmarks and mapped to the 16-qubit device *IBM QX5*. Besides that, benchmarks that are relevant for existing quantum algorithms such as quantum ripple-carry adders (based on the realization proposed in [36] and denoted *adder*) and small versions of Shor's algorithm (based on the realization proposed in [13] and denoted *shor*) have been considered. The evaluation has been conducted on a 4.2 GHz machine with 4 cores (2 hardware threads each) and 32 GB RAM.

17.3.1 Effect of the Optimizations

The first part of the evaluation determines the improvements gained by the optimizations discussed in Sect. 17.2. The corresponding numbers are listed in Table 17.1. For each benchmark, Table 17.1 provides the name, the number of logical qubits n, and the number of gates $|G|$ before mapping the circuit to IBM QX5. The remainder of the table lists the results provided by the presented methodology, i.e., the number of gates of the circuit $|G|$ after mapping it to IBM QX5 as well as the time required to determine that mapping (in CPU seconds).

Three different settings of the methodology are thereby considered. As baseline serves the approach presented in Sect. 17.1 that uses an A^* algorithm to determine locally optimal mappings for each layer of the circuit (denoted *Baseline* in the following). Furthermore, the numbers are listed when enriching the baseline with a look-ahead scheme as discussed in Sect. 17.2 (denoted *Look-Ahead* in the following). Finally, the resulting numbers for the fully optimized methodology that uses a look-ahead scheme and additionally allows for evolving the mapping throughout the mapping process as discussed in Sect. 17.2 (denoted *Fully-Optimized* in the following) are listed. The timeout was set to 1 h.

Table 17.1 clearly shows the improvements that are gained by applying the optimizations discussed in Sect. 17.2. On average, the number of gates of the mapped circuit decreases by 16.1% when applying a look-ahead scheme as discussed in Sect. 17.2. However, using the look-ahead scheme causes the mapping algorithm to timeout in nine cases (instead of five cases for *baseline*). If the initial mapping of logical qubits to physical qubits is additionally evolved throughout the mapping process instead of starting with a random mapping, this scalability issue is overcome while obtaining mappings of similar quality. In fact, the average improvement regarding the number of gates of the circuits slightly increases to 19.7% (compared to *Baseline*).

Table 17.1 Effect of the optimizations

Name	n	$|G|$	Baseline $|G|$	t	Look-ahead $|G|$	t	Fully-optimized $|G|$	t
adder 10	10	142	444	1.22	355	1.08	292	1.20
hwb9	10	207,775	743,973	1662.82	653,249	1437.33	655,220	1422.33
ising_model_10	10	480	235	4.63	235	4.58	251	4.14
max46	10	27,126	–	TO	86,049	185.19	84,914	185.82
mini_alu	10	173	710	2.38	587	1.32	474	1.25
qft_10	10	200	685	1.23	445	1.54	447	1.25
rd73	10	230	952	1.83	916	1.57	656	1.52
sqn	10	10,223	37,781	80.15	32,099	72.25	32,095	68.97
sym9	10	21,504	78,388	172.95	67,290	147.99	66,637	145.37
sys6-v0	10	215	962	1.96	794	1.50	613	1.36
urf3	10	125,362	517,104	1045.85	439,268	888.77	440,509	873.84
9symml	11	34,881	133,813	296.80	114,179	255.02	116,508	254.25
dc1	11	1914	8310	16.22	6024	13.07	5946	12.38
life	11	22,445	86,075	358.56	73,020	161.90	74,632	166.95
shor_11	11	49,295	125,825	325.13	109,574	317.81	106,322	322.78
sym9	11	34,881	133,813	292.40	114,179	249.49	116,508	251.42
urf4	11	512,064	1,926,128	4257.19	1,653,689	3481.55	1,650,845	3534.79
wim	11	986	3632	7.60	3176	6.54	2985	6.30
z4	11	3073	12,041	24.91	10,002	20.71	9717	20.92
adder_12	12	177	631	1.76	483	1.64	372	1.32
cm152a	12	1221	4254	9.13	4039	8.23	3738	8.02
cycle10_2	12	6050	23,991	49.62	19,513	45.82	19,857	42.26
rd84	12	13,658	52,508	157.72	45,509	107.69	45,497	99.89
sqrt8	12	3009	11,921	26.64	10,166	21.35	9744	19.66
sym10	12	64,283	251,731	535.84	214,881	500.05	215,569	501.02
sym9	12	328	1436	2.54	1240	2.27	955	2.08
adr4	13	3439	13,475	29.53	11,245	23.84	11,301	23.17
dist	13	38,046	147,115	323.20	125,342	334.67	125,867	291.90
gse_10	13	390,180	863,511	2441.38	576,399	2263.71	520,010	2237.10
ising_model_13	13	633	313	6.08	313	6.09	329	5.11
plus63mod4096	13	128,744	529,896	1203.45	434,900	1006.16	439,981	1086.48
radd	13	3213	11,790	25.35	10,868	23.67	10,441	22.00
rd53	13	275	1367	2.53	1044	1.97	942	1.93
root	13	17,159	67,941	327.20	56,654	120.01	57,874	120.82
shor_13	13	98,109	259,511	656.43	229,752	783.74	224,556	640.55
squar5	13	1993	7948	16.35	6453	13.29	6267	12.96
410,184	14	211	914	1.82	708	1.42	758	1.48
adder_14	14	212	–	TO	–	TO	437	1.47
clip	14	33,827	135,455	322.36	–	TO	114,336	327.55
cm42a	14	1776	6473	13.93	5572	11.16	5431	11.95

(continued)

Table 17.1 (continued)

Name	n	|G|	Baseline		Look-ahead		Fully-optimized	
			|G|	t	|G|	t	|G|	t
cm85a	14	11,414	46,300	185.98	37,927	464.90	37,746	242.80
plus127mod8192	14	330,777	–	TO	–	TO	1,132,251	2481.95
plus63mod8192	14	187,112	773,514	1628.19	637,137	1364.63	640,204	1443.33
pm1	14	1776	6473	13.62	5572	11.14	5431	11.10
sao2	14	38,577	155,351	330.62	–	TO	131,002	283.90
sym6	14	270	1101	2.33	1136	2.05	852	1.84
co14	15	17,936	80,399	331.89	62,348	176.55	63,826	133.71
dc2	15	9462	36,968	83.96	31,722	95.81	30,680	72.53
ham15	15	8763	32,175	79.80	27,861	61.70	28,310	68.75
misex1	15	4813	17,833	38.63	15,260	33.18	15,185	33.11
rd84	15	343	1593	3.30	1337	2.81	971	2.23
square_root	15	7630	–	TO	–	TO	25,212	55.35
urf6	15	171,840	684,701	1456.37	–	TO	580,295	1436.16
adder_16	16	247	–	TO	–	TO	515	1.72
alu2	16	28,492	118,919	244.83	–	TO	98,166	454.93
cnt3-5	16	485	1957	3.89	1488	2.98	1376	3.00
example2	16	28,492	118,919	246.00	–	TO	98,166	449.08
inc	16	10,619	41,042	86.91	34,742	74.13	34,375	72.85
ising_model_16	16	786	391	6.88	391	6.86	426	6.47
qft_16	16	512	2193	69.04	1299	8.62	1341	16.43

n: the number of qubits
$|G|$: the number of quantum gates (elementary operations)
d: depth of the quantum circuits
t: run-time of the algorithm
Baseline: the approach described in Sect. 17.1.2
Look-Ahead: the approach described in Sect. 17.1.2 enriched with the look-ahead scheme discussed in Sect. 17.2
Fully-Optimized: the approach described in Sect. 17.1.2 enriched with all optimizations discussed in Sect. 17.2 The timeout was set to 1 h

Overall, the optimizations discussed in Sect. 17.2 not only increase the scalability of the mapping algorithm outlined in Sect. 17.1.2, but—as a positive side effect—also reduce the size of the resulting circuit.

17.3.2 Comparison to IBM's Approach

The second part of the evaluation compares the presented heuristic mapping methodology to the solution provided by IBM via Qiskit 0.4.15. A fair comparison of both mapping solution is guaranteed since the discussed mapping algorithm has been integrated into *Qiskit*. Hence, the same decomposition schemes as well as the same post-mapping optimizations are applied in both cases.

Table 17.2 lists the respectively obtained results. For each benchmark, Table 17.2 lists the name, the number of logical qubits n, the number of gates $|G|$, and the depth d of the quantum circuit before mapping it to IBM QX5. The remaining columns list the number of gates, the depth, and the run-time t (in CPU seconds)

Table 17.2 Mapping to IBM QX5

Name	n	$|G|$	d	IBM's solution			Presented heuristic approach		
				$|G|_{min}$	d_{min}	t_{min}	$|G|$	d	t
adder˙10	10	142	99	382	203	5.75	292	172	1.20
hwb9	10	207,775	116,199	–	–	TO	655,220	375,105	1422.33
ising_model_10	10	480	70	347	73	6.23	251	47	4.14
max46	10	27,126	14,257	105,651	53,397	1652.31	84,914	46,270	185.82
mini_alu	10	173	69	707	290	9.82	474	225	1.25
qft_10	10	200	63	670	210	9.49	447	170	1.25
rd73	10	230	92	930	393	12.74	656	301	1.52
sqn	10	10,223	5458	39,175	20,329	627.32	32,095	17,801	68.97
sym9	10	21,504	12,087	80,867	43,707	1314.22	66,637	38,849	145.37
sys6-v0	10	215	75	853	329	11.29	613	250	1.36
urf3	10	125,362	70,702	–	–	TO	440,509	239,702	873.84
9symml	11	34,881	19,235	143,042	73,363	2233.67	116,508	64,279	254.25
dc1	11	1914	1038	7283	3859	113.8	5946	3378	12.38
life	11	22,445	12,511	91,724	47,471	1446.67	74,632	41,767	166.95
shor_11	11	49,295	30,520	124,160	67,962	2149.13	106,322	58,943	322.78
sym9	11	34,881	19,235	142,431	72,959	2237.2	116,508	64,279	251.42
urf4	11	512,064	264,330	–	–	TO	1,650,845	878,249	3534.79
wim	11	986	514	3834	1947	59.01	2985	1711	6.30
z4	11	3073	1644	11,905	6024	188.53	9717	5335	20.92
adder_12	12	177	123	579	279	8.81	372	226	1.32
cm152a	12	1221	684	4761	2501	74.61	3738	2155	8.02
cycle10_2	12	6050	3386	25,362	13,125	392.12	19,857	11,141	42.26
rd84	12	13,658	7261	56,134	28,172	860.9	45,497	24,473	99.89
sqrt8	12	3009	1659	12,541	6398	194.8	9744	5501	19.66
sym10	12	64,283	35,572	–	–	TO	215,569	118,753	501.02
sym9	12	328	127	1411	582	20.09	955	425	2.08
adr4	13	3439	1839	13,638	6991	210.34	11,301	6205	23.17
dist	13	38,046	19,694	158,516	77,027	2412.78	125,867	66,318	291.90
gse_10	13	390,180	245,614	–	–	TO	520,010	376,695	2237.10
ising_model_13	13	633	71	439	82	7.87	329	47	5.11
plus63mod4096	13	128,744	72,246	–	–	TO	439,981	243,861	1086.48
radd	13	3213	1781	12,674	6716	206.15	10,441	5872	22.00
rd53	13	275	124	1223	518	16.65	942	469	1.93
root	13	17,159	8835	71,721	34,798	1094.63	57,874	30,068	120.82

(continued)

Table 17.2 (continued)

Name	n	$\|G\|$	d	IBM's solution			Presented heuristic approach		
				$\|G\|_{min}$	d_{min}	t_{min}	$\|G\|$	d	t
shor_13	13	98,109	59,350	–	–	TO	224,556	118,536	640.55
squar5	13	1993	1049	8111	4073	124.09	6267	3448	12.96
410,184	14	211	104	864	393	13.36	758	366	1.48
adder_14	14	212	147	659	332	9.76	437	268	1.47
clip	14	33,827	17,879	144,737	70,732	2197.11	114,336	60,882	327.55
cm42a	14	1776	940	6623	3480	104.23	5431	3013	11.95
cm85a	14	11,414	6374	47,908	24,798	742.46	37,746	21,189	242.80
plus127mod8192	14	330,777	185,853	–	–	TO	1,132,251	626,451	2481.95
plus63mod8192	14	187,112	105,142	–	–	TO	640,204	354,076	1443.33
pm1	14	1776	940	6488	3444	104.21	5431	3013	11.10
sao2	14	38,577	19,563	163,679	77,525	2495.54	131,002	66,975	283.90
sym6	14	270	135	1092	514	16.05	852	456	1.84
co14	15	17,936	8570	83,301	35,926	1177.71	63,826	30,366	133.71
dc2	15	9462	5242	38,807	20,155	601.92	30,680	17,269	72.53
ham15	15	8763	4819	35,150	18,293	546.05	28,310	15,891	68.75
misex1	15	4813	2676	19,090	10,172	299.9	15,185	8729	33.11
rd84	15	343	110	1579	529	19.5	971	353	2.23
square_root	15	7630	3847	30,349	14,828	461.14	25,212	13,205	55.35
urf6	15	171,840	93,645	–	–	TO	580,295	313,011	1436.16
adder_16	16	247	171	968	437	13.88	515	319	1.72
alu2	16	28,492	15,176	125,601	60,839	1905.14	98,166	51,817	454.93
cnt3-5	16	485	209	1899	825	27.08	1376	669	3.00
example2	16	28,492	15,176	125,022	60,543	1900 17	98,166	51,817	449.08
inc	16	10,619	5863	43,097	22,413	672.39	34,375	19,176	72.85
ising_model_16	16	786	71	785	149	12.44	426	48	6.47
qft_16	16	512	105	2056	521	25.6	1341	404	16.43

n: the number of qubits
$\|G\|$: the number of quantum gates
d: depth of the quantum circuit
t: run-time of the algorithm
For IBM's solution, the obtained minimum of 5 runs is listed. The timeout was set to 1 h

for IBM's solution as well as for the heuristic approach presented in this chapter. Since IBM's mapping algorithm searches for mappings that satisfy all coupling-constraints randomly (guided by certain heuristics), the mapping procedure was conducted 5 times for each benchmark and the obtained minimum is listed (denoted by subscript $_{min}$). The timeout for searching a single mapping was again set to 1 h.

The results clearly show that the presented heuristic solution can efficiently tackle the considered mapping problem—in particular compared to the method available thus far. While IBM's solution runs into the timeout of 1 h in 10 out of 60 cases, the algorithm presented in this chapter determines a mapping for each circuit within the given time limit. Besides that, the approach is frequently one order of magnitude faster compared to IBM's solution.

Besides efficiency, the presented methodology for mapping a quantum circuit to the IBM Q devices also yields circuits with significantly fewer gates than the results determined by IBM's solution. In fact, the solution discussed in Sect. 17.1 results on average in circuits with 24.0% fewer gates and 18.3% fewer depth on average compared to the minimum observed when running IBM's algorithm several times.

Chapter 18
A Dedicated Heuristic Approach for SU(4) Quantum Circuits

Abstract This chapter provides a more elaborated approach for mapping a certain kind of random circuits (introduced by IBM in context of their *Qiskit Developer Challenge*), which is composed of a pre-processing step to reduce the complexity beforehand, an actual mapping step, as well as dedicated post-processing that reduces the costs of the mapped circuit. The resulting solution was declared winner of the challenge mentioned above because it consistently produces circuits with at least 10% better cost than the competition and is more than 6 times faster than all the others.

Keywords Quantum circuits · Quantum gates · mapping · A* search · SWAP gates · IBM Q · Coupling map · Qubits

Recently a set of certain random quantum circuits (called $SU(4)$ *quantum circuits* in the following) has been introduced for validating quantum computers [34]. Moreover, these circuits have explicitly been advocated by IBM to benchmark compilers (and, hence, mapping algorithms) through the *Qiskit Developer Challenge* [80]. However, it turns out that $SU(4)$ quantum circuits constitute a worst case for the heuristic method discussed in the previous chapter—making it infeasible. Hence, for a class of circuits, which is considered to be important by a major player in the development of quantum computers, no method exists for efficiently compiling/mapping them, e.g., to IBM Q devices.

This chapter (based on [199]) addresses this problem by providing a *dedicated* approach for mapping $SU(4)$ quantum circuits to IBM Q devices. To this end, the heuristic approach introduced in Chap. 17 is reviewed to determine its benefits and bottlenecks. Based on that evaluation, a dedicated approach is presented which explicitly takes the structure of $SU(4)$ quantum circuits into consideration. The conducted evaluation shows that this approach significantly outperforms IBM's current solution with respect to fidelity of the resulting circuits as well as regarding run-time. Moreover, it has been declared winner of the IBM Qiskit Developer

© Springer Nature Switzerland AG 2020
A. Zulehner, R. Wille, *Introducing Design Automation for Quantum Computing*,
https://doi.org/10.1007/978-3-030-41753-6_18

Challenge since—according to IBM—it yields compiled circuits with at least 10% better costs than the other submissions while generating them at least 6 times faster.

In the following, the dedicated approach for $SU(4)$ quantum circuits is presented. To this end, Sect. 18.1 describes the considered circuits. Based on that, Sect. 18.2 discusses the dedicated approach that utilizes a pre-processing step to reduce the complexity beforehand, a powerful mapping strategy based on A* search, as well as a dedicated post-mapping optimization to reduce the costs of the mapped circuits. Eventually, the presented approach is evaluated in Sect. 18.3—including a comparison to IBM's own mapping solution available in Qiskit.

18.1 Considered Circuits

This section describes the random quantum circuits utilized for validating quantum circuits [34]. These circuits are also provided by IBM to benchmark the performance of respective compilers through the *Qiskit Developer Challenge* [80]. The circuits are products of random two-qubit gates from $SU(4)$ that are applied to random pairs of qubits and denoted as $SU(4)$ *quantum circuits* in the following.[1] More precisely, in each layer of the circuit, the available qubits are grouped randomly into pairs of two qubits each (if their number is even). Then, to each of these pairs of qubits, a random two-qubit gate from $SU(4)$ is applied. Since these two-qubit gates are not available in the gate set of the IBM QX devices, KAK-decomposition [160] is used to decompose each of these two-qubit gates into a sequence of (up to) three CNOTs and seven one-qubit gates. Eventually, these decomposed gates form the circuits for determining the performance of the compilers/mapping algorithms.

Example 18.1 *Figure 18.1 shows the KAK-decomposition of a random $SU(4)$ gate. For simpler visualization, the parameters θ, ϕ, and λ are neglected for the one-qubit gates U^i (which are usually different for each U^i). As can be seen, one-qubit gates and CNOT gates are applied in an interleaved fashion.*

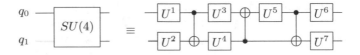

Fig. 18.1 KAK-decomposition of an $SU(4)$ gate

---

[1] $SU(4)$ is the *special unitary group with degree 4*, i.e., the Lie group of 4×4 unitary matrices with determinant 1. The functionality of any two-qubit gate is described by an element from this group.

18.2 Proposed Approach

This section describes the proposed dedicated procedure to compile/map $SU(4)$ quantum circuits to IBM Q devices. Such a solution is required since the heuristic approach discussed in Chap. 17 is hardly suitable for the $SU(4)$ circuits reviewed in Sect. 18.1, because:

- The solution rests on the main idea to first divide the circuit into layers of gates[2] and, afterwards, determines a permutation of qubits for each layer which satisfies all coupling-constraints within this subset of gates.[3]
- $SU(4)$ circuits are composed of layers of gates which frequently contain $\frac{n}{2}$ different CNOT configurations (with n being the number of qubits). This is a worst-case scenario since the more CNOT gates are employed within a layer, the more constraints have to be satisfied by a permutation of the qubits.

As a consequence, the solution presented in Chap. 17 cannot unfold its power for determining mapped circuits with smaller overhead than IBM's solution when applied for $SU(4)$ circuits as it basically has to check all permutations within a layer until one is determined satisfying all constraints imposed by the CNOTs. Considering that $SU(4)$ circuits have explicitly been provided by IBM to benchmark compilers, this is a serious drawback and motivates a dedicated approach for this kind of circuits. To overcome the limitations while keeping the availability of a look-ahead scheme, each gate is considered on its own (instead of using a layered-based approach). In order to deal with the correspondingly resulting complexity, the proposed algorithm employs a combination of three steps: a pre-process step (reducing the complexity beforehand), a powerful search method (solving the mapping problem), and eventually a dedicated post-mapping optimization (exploiting further optimization potential after the mapping).

18.2.1 Pre-Processing: Grouping Gates

Since each gate is considered on its own, the mapping may change after each gate (requiring many calls of the mapping algorithm). To reduce the number of possible changes, a pre-processing step is performed where groups of gates are formed, which are represented as a directed acyclic graph (DAG).[4] By this, the mapping algorithm has to be called (at most) only once per group instead of once per gate. As further advantage, this DAG representation inherently encodes the precedence

---

[2]A layer contains only gates that act on disjoint qubits. Thus, all gates of a layer can be applied in parallel.

[3]Between the respective layers, SWAP gates as shown in Fig. 15.1 on Page 162 are introduced to establish the respective qubit permutations.

[4]This may also be called a dependency graph.

of the groups of gates and, thus, unveils important information about which groups of gates commute—giving the degree of freedom to choose which group shall be mapped next.

In order to group the gates, the circuit is topologically sorted before grouping all gates that act on pairs of logical qubits (e.g., on qubits q_i and q_j) into a group G_k. This includes one-qubit gates on q_i or q_j as well as CNOTs with control q_i and target q_j (or vice versa). This grouping is done in a greedy fashion—until observing a CNOT with control or target q_i (q_j) that acts on a qubit different from q_j (q_i). This is possible, since gates that act on distinct sets of qubits are commutative.

Example 18.2 *Consider again the circuit shown at the right-hand side of Fig. 18.1. Since, all gates of the circuit act on qubits q_0 and q_1, the grouped circuit contains a single group. By this, the mapping has to be changed at most once in order to apply all gates.*

As stated above, grouping gates has a positive effect on the following mapping algorithm, since all gates of a group can be applied once the coupling-constraints are satisfied for the involved qubits.[5] Thus, the mapping of the gates of the circuit reduces to mapping the groups.

Example 18.3 *Consider the DAG shown in Fig. 18.2. This DAG represents a quantum circuit composed of 6 qubits, where the first layer is composed of SU(4) gates between the logical qubits q_0 and q_1, q_2 and q_3, as well as q_4 and q_5, respectively. Moreover, the second layer contains SU(4) gates between the logical qubits q_1 and q_2, q_3 and q_4, as well as q_0 and q_5, respectively.*

18.2.2   Solving the Mapping Problem

After grouping the gates, the coupling-constraints of the target architecture given by the coupling map are satisfied by a mapping algorithm that determines a dynamically changing mapping of the logical qubits to the physical ones. In theory, the mapping can change (by inserting SWAP gates) after each group (cf. Chap. 16)—resulting in a huge search space since $m!$ possibilities exist for each

Fig. 18.2 DAG after grouping the gates of the circuit

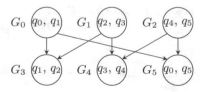

---

[5]Note that the direction of the CNOTs might have to be adjusted (which is rather cheap since only Hadamard gates have to be added).

such intermediate mapping. To cope with this enormous search space an A* search algorithm (cf. Sect. 3.2) is utilized to find a solution that is as cheap as possible.

For the mapping strategy presented in this chapter, an arbitrary initial mapping is chosen such that the coupling-constraints are satisfied (i.e., the corresponding logical qubits are mapped to physical ones that are connected in the coupling map) for all groups in the DAG that do not have any predecessors. By this, the gates of these groups can be immediately added to the (initially empty) mapped/compiled circuit.[6]

Example 18.4 *Consider again the DAG in Fig. 18.2, which describes the gate groups to be mapped. Assume that the circuit shall be mapped to IBM QX5, whose coupling map is depicted in Fig. 2.2d on Page 17. One possible initial mapping is $q_0 \rightarrow Q_1$, $q_1 \rightarrow Q_0$, $q_4 \rightarrow Q_2$, $q_2 \rightarrow Q_{15}$, $q_5 \rightarrow Q_3$, and $q_3 \rightarrow Q_{14}$ (i.e., the logical qubits are mapped to the six leftmost physical qubits). Using this initial mapping, the gate groups in the first layer (i.e., G_0, G_1, and G_2) can be applied since the involved logical qubits are mapped to physical ones that are connected in the coupling map for each of the groups.*

After determining an initial mapping, the actual mapping procedure is composed of two alternating steps that are employed until all groups are mapped.

The first step adds all groups to the compiled circuit, whose parents in the DAG have already been mapped and whose logical qubits are mapped to physical ones that are connected in the coupling map.

Example 18.4 (continued) *The initial mapping additionally allows adding gates of group G_3 to the compiled circuit, since its parents in the DAG (i.e., the groups G_1 and G_2) have already been mapped and the coupling-constraints are also satisfied (since $q_1 \rightarrow Q_0$ and $q_2 \rightarrow Q_{15}$).*

The second step determines the set of groups G_{next} that can be applied next according to their precedence in the circuit, i.e., the set of groups whose parents in the DAG are already compiled. Then, the task of the mapping algorithm is to determine a new mapping (by inserting SWAP gates) such that the coupling-constraints are satisfied for at least one of the groups in G_{next}.

Example 18.4 (continued) *One possibility is to incorporate a SWAP operation on the physical qubits Q_{15} and Q_2 since this "moves" the logical qubits q_3 and q_4 towards each other and, thus, allows adding the gates from gate group G_4 to the compiled circuit. Finally, inserting another SWAP operation between the physical qubits Q_1 and Q_2 allows adding the gates of the group G_5 to the compiled circuit. Overall, two SWAP gates were inserted during the mapping procedure of the circuit.*

Another solution would be to incorporate a SWAP operation on the physical qubits Q_2 and Q_3. Since this "moves" the logical qubits q_0 and q_5, as well as the

---

[6]Note that the qubits have to be relabeled according to the mapping and that the direction of some CNOTs might be adjusted.

logical qubits q_3 and q_4 towards each other, the gate groups G_4 and G_5 can be applied by inserting a single SWAP operation during the compilation procedure.

Among the solutions found by the mapping algorithm, the mapping that yields the lowest costs is desired. Since there are $m!$ different mappings of the physical qubits, an A^* search is utilized to avoid exploring the whole search space. The general description of the A^* search algorithm (cf. Sect. 3.2) has been adjusted for the considered mapping problem. More precisely, the *root node* represents the current mapping of the logical qubits to the physical ones. A *goal node* represents any mapping where the coupling-constraints are satisfied for at least one of the groups. *Expanding* a node is conducted by applying one SWAP operation between two physical qubits, which results in a successor mapping. Given that, the corresponding cost functions $f(x)$ and $h(x)$ have to be determined. The fixed costs $f(x)$ of a node is given by the number of SWAP operations that have been added (starting from the current mapping). For the estimation of the remaining costs $h(x)$, the utilized heuristic employs a look-ahead scheme, which allows significantly reducing the costs of the compiled circuit.

More precisely, for each group, the distance of the physical qubits (considering the coupling map) where the respective logical qubits are mapped to is determined, before summing up these distances up for all groups in G_{next}.[7] By this, the focus is not on one of these groups, the proposed approach rather tries to optimize the mapping for groups that are applied in the near future.

Example 18.4 (continued) *The look-ahead scheme determines the goal node reached by conducting a SWAP operation between the physical qubits Q_2 and Q_3, since from the two solutions resulting in a goal node with cost 1 (inserting a single SWAP gate; as discussed above), the solution with the lower look-ahead costs was chosen.*

18.2.3   Post-Mapping Optimizations

After satisfying the coupling-constraints given by the target architecture, a dedicated post-mapping optimization is applied in order to further reduce the costs of the compiled circuit. To this end, the gates of the compiled circuit are regrouped as described in Sect. 18.2.1, since the mapping algorithm has added several SWAP gates to the compiled circuit. Then, the resulting DAG is traversed while optimizing each group individually.

The key idea of the proposed optimization is that the functionality of the gates in a group G_i is represented by a single matrix from $SU(4)$. Hence, building up this

---

[7]Note that the heuristic is not admissible and, hence, may not lead to a locally optimal solution. However, locally optima are not desired anyways, since these often yield a globally larger overhead.

matrix by multiplying the unitary matrices representing the individual gates and, again, using KAK-decomposition [160] allows determining another group G'_i with 3 CNOTs and 7 one-qubit gates that realizes the same functionality (cf. Sect. 18.1). If the gates in G'_i have lower costs than the gates in the original group G_i, G'_i replaces G_i in the DAG. This especially works well, when applying a SWAP gate to two qubits, to which a gate from $SU(4)$ has been applied right before.

Example 18.5 *Consider again the KAK-decomposition shown in Fig. 18.1 with its 3 CNOTs and 7 one-qubit gates. Furthermore, assume that immediately afterwards a SWAP operation is applied to the physical qubits currently holding the logical qubits q_0 and q_1—yielding a group G_i with 6 CNOTs and 11 one-qubit gates. However, representing the overall functionality of this group as a unitary matrix from $SU(4)$ and applying KAK-decomposition again yields another group G'_i with, again, 3 CNOTs and 7 one-qubit gates. Hence, the SWAP operation is conducted "for free."*

Note that the knowledge of this post-mapping optimization can be used to improve the mapping algorithm itself. More precisely, knowing that SWAP operations directly applied after a gate from $SU(4)$ are "for free" is included in the costs function $f(x)$ of the fixed costs by setting the costs of the respective SWAP operation to 0.

Finally, a similar (but simpler) optimization is applied for optimizing subsequent one-qubit gates within a group. Such gates may occur, e.g., when swapping the direction of a CNOT by inserting Hadamard gates. Again, the 2×2 unitary matrices describing the individual gates can be multiplied. Afterwards, the Euler angles of the rotations around the z- and y-axis are determined.

Example 18.6 *Consider again the KAK-decomposition shown in Fig. 18.1. To change the direction of the center CNOT gate, Hadamard gates are inserted to each qubit before and after the CNOT—yielding to subsequent one-qubit gates that are applied to q_0 and q_1, respectively. Again, this sequence of, e.g., U_3 and H can again be represented by a single one-qubit gate.*

18.3 Evaluation

This section evaluates the presented dedicated approach and compares it to the IBM's approach available in Qiskit 0.4.15 [33].[8] To this end, it has been implemented in Cython (available at http://iic.jku.at/eda/research/ibm_qx_mapping) and the scripts provided by IBM have been used to conduct the evaluation (these scripts are available at [80]). Since the fidelity of CNOT gates is approximately 10 times lower than the fidelity of one-qubit gates on IBM Q devices (cf. [79]), the provided

---

[8]Note that no comparison is reported for the approach presented in Chap. 17 since the $SU(4)$ circuits represent a worst case for them and, hence, this method is not feasible for these benchmarks.

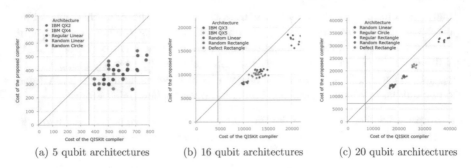

(a) 5 qubit architectures   (b) 16 qubit architectures   (c) 20 qubit architectures

Fig. 18.3 Costs of the compiled circuits

cost function assigns a cost of 10 for each CNOT as well as a cost of 1 for one-qubit gate.[9] The evaluation has been conducted on a 3.8 GHz machine with 32 GB RAM.

Besides the circuits, IBM also provides several coupling maps for devices with 5, 16, and 20 qubits, respectively. These devices include the existing quantum devices *IBM QX2*, *IBM QX4*, and *IBM QX5*, as well as others where the qubits are arranged in a linear, circular, or rectangular fashion. For these devices, the direction of the arrows in the coupling maps are chosen randomly by IBM (or connections are missing at all) to provide a realistic basis for the evaluation. For each number of qubits in the devices (5, 16, or 20), 10 circuits are utilized, which are compiled to each architecture with the corresponding number of qubits. Eventually, this results in a setting which is also used by IBM to evaluate compilers submitted to the Qiskit Developer Challenge [80]. The resulting costs are visualized by means of scatter plots in Fig. 18.3.

Each plot in Fig. 18.3 shows the costs of the compiled circuits when using the Qiskit compiler on the x-axis, as well as the costs of the compiled circuit when using the proposed solution on the y-axis. Each point represents one $SU(4)$ circuit that is compiled for a certain architecture. Hence, a point underneath the main diagonal indicates that the proposed solution yields a circuit with lower costs (which is the case for all evaluated circuits and architectures). The larger the distance to the main diagonal, the larger the improvement. Moreover, horizontal and vertical lines indicate the costs of the original circuits (i.e., the costs before compilation).

As can be seen in Fig. 18.3a, circuits compiled by the dedicated methodology may be cheaper than the original circuit (despite the fact that SWAP gates are *added* during the compilation process). This is possible since, in some cases, two $SU(4)$ gates are subsequently applied to the same two qubits. By using the discussed post-mapping optimization (cf. Sect. 18.2.3), these gates are combined to a single two-qubit gate from $SU(4)$. Overall, an average improvement by a factor of 1.54

---

[9]Note that a one-qubit gate $U(0, 0, \lambda)$ has cost 0 since no pulse is applied to the respective qubit in this case.

Table 18.1 Average improvement factors

5 qubits			16 qubits			20 qubits		
Architecture	Costs	Time	Architecture	Costs	Time	Architecture	Costs	Time
IBM QX2	1.55	5.96	IBM QX3	1.25	14.85	Random linear	1.15	14.64
IBM QX4	1.54	5.84	IBM QX5	1.23	12.87	Regular circle	1.19	14.25
Regular linear	1.58	5.40	Random linear	1.19	11.52	Regular rectangle	1.24	32.67
Random linear	1.57	5.43	Random rectangle	1.24	20.99	Random rectangle	1.27	33.95
Random circle	1.49	5.81	Defect rectangle	1.39	25.80	Defect rectangle	1.25	21.78
Avg	**1.54**	**5.68**	**Avg**	**1.26**	**16.42**	**Avg**	**1.22**	**21.90**

compared to IBM's own solution is observed for the 5-qubit architectures. For the 16- and 20-qubit architectures, the probability that two subsequent $SU(4)$ gates are applied to the same qubits is almost zero. Although this does not allow as much post-mapping optimization as for the 5-qubit architectures, significant improvements of a factor of 1.26 and 1.22 on average are still observed, respectively. The precise improvements for each architecture are listed in Table 18.1.

Besides the average improvement in terms of the provided cost function, the presented dedicated method is also significantly faster than IBM's solution. While IBM's solution requires more than 200 s for mapping some of the circuits composed of 20 qubits, the presented method was able to map each of the circuits within 10 s. On average, an improvement of the run-time by a factor of 5.68, 16.42, and 21.90 for the architectures with 5, 16, and 20 qubits is obtained, respectively (cf. Table 18.1).

Overall, the evaluation using the scripts, circuits, and coupling maps provided by IBM shows that the dedicated compile methodology presented in this chapter significantly outperforms IBM's own solution regarding the provided cost function (which estimates fidelity) as well as run-time. Moreover, the solution proposed in this chapter has been declared winner of the Qiskit Developer Challenge. According to IBM, it yields compiled circuits with at least 10% better costs than the other submissions while generating them at least 6 times faster.

Chapter 19
Summary

Abstract This chapter summarizes Part IV of this book by reflecting the presented strategies for mapping quantum circuits to NISQ devices (considering IBM Q devices as representatives). Moreover, the gained improvements compared to the state of the art are also discussed.

Keywords Quantum circuits · Quantum gates · Mapping · IBM Q · Coupling map · Qubits

This part of the book presented several efficient methods for mapping quantum circuits to the IBM Q devices—an $\mathcal{NP}$-complete problem. Even though the approaches have been developed for IBM's approach, they can be easily adjusted to work for other NISQ devices as well—by formulating their coupling-constraints symbolically (when aiming for an exact approach) or within the search algorithm (when aiming for a heuristic solution). All presented methods are publicly available at http://iic.jku.at/eda/research/ibm_qx_mapping.

Chapter 16 presented the first exact approach that is capable of mapping quantum circuit to the IBM Q devices with minimal overhead (by means of added SWAP and H operations). This does not only give an optimal mapping solution, but can also be utilized to evaluate how far existing heuristic solutions exceed this lower bound. In fact, the evaluation shows that the overhead generated by IBM's own mapping solution exceeds the lower bound by more than 100% on average—and that already for rather small circuits to be mapped. This unveiled potential for improvement further motivates the development of new heuristic approaches (since exact ones are obviously not feasible for large instances due to complexity of the underlying problem).

To come closer to an optimal solution while remaining scalable, Chap. 17 presented one of the first heuristic approaches that shows significant improvements regarding gate count, depth, and run-time—clearly outperforming IBM's solution, e.g., on a 16-qubit architectures and for circuits composed of thousands of gates. This difference in quality is mainly because IBM's solution (which is based in Bravyi's algorithm and available at [33]) randomly searches for a mapping that

© Springer Nature Switzerland AG 2020   201
A. Zulehner, R. Wille, *Introducing Design Automation for Quantum Computing*,
https://doi.org/10.1007/978-3-030-41753-6_19

satisfies the coupling-constraints—leading to a rather small exploration of the search space. In contrast, the presented approach aims for utilizing algorithms known from design automation of conventional circuits and systems to explore a more suitable part of the search space and to additionally exploiting information of the circuit. More precisely, a look-ahead scheme is employed that considers gates that are applied in the near future and, thus, allows determining mappings which aim for a global minimum (with respect to the number of SWAP and H operations). Besides the gained improvements over IBM's solution, the presented heuristic approach has also triggered interest of several research groups on mapping of quantum circuits to IBM Q devices.

Finally, a dedicated solution focusing on the so-called $SU(4)$ quantum circuits (utilized for validating quantum computers [34]) has been presented in Chap. 18. These circuits have also been considered in context of the *IBM Qiskit Developer Challenge* [80] to benchmark the performance of the compilers. Since these circuits are a worst-case scenario for the heuristic method described in Chap. 17 (making this approach infeasible for these specific circuits), a dedicated approach that explicitly takes the structure of these circuits into account is required. By using a three stage approach composed of pre-processing (reducing the complexity beforehand), a powerful search algorithm to solve the mapping task, as well as dedicated post-mapping optimization (reducing the number of required gates), IBM's mapping approach is significantly outperformed regarding run-time and the fidelity of the mapped circuit. Moreover, the presented approach has been declared as winner of the *IBM Qiskit Developer Challenge* since it constantly yields circuits with 10% lower costs than the other submissions while generating them at least 6 times faster (according to IBM). This also triggered an integration into IBM's Qiskit and Atos' QLM.

Part V
Conclusion

Chapter 20
Conclusion

Abstract This chapter concludes the book by summarizing the considered design tasks for quantum computing, as well as the presented methods that often outperform the current state of the art significantly (by orders of magnitudes in many cases). This again demonstrates the importance of introducing knowledge from the design automation community into the quantum realm, and showcases the available potential.

Keywords Quantum computing · Design automation · Quantum circuits · Simulation · Synthesis · Mapping

This book presented substantial contributions to the recently starting (and highly demanded) trend of bringing knowledge from design automation for conventional system to quantum computing—trying to avoid situations where the unavailability of efficient design tools hinders the usage of developed hardware following this powerful computing paradigm. In fact, investigating the important design tasks *quantum-circuit simulation*, the *design of occurring Boolean components*, as well as the *mapping task* (the latter two are part of the compilation flow required to run quantum algorithms on real hardware) from a design automation perspective leads to a variety of tools inspired from clever as well as efficient algorithms and data structures that have been discussed in this book. In many cases, improvements of several orders of magnitudes regarding run-time or other design objectives have been achieved. These achievements are summarized as follows.

A complementary *quantum-circuit simulation* approach has been presented that utilizes decision diagrams to represent occurring quantum states and operations much more compactly—overcoming the exponential overhead that the current state of the art is currently facing in many cases. More precisely, this often leads to situations where simulation times are reduced from days and hours to minutes and seconds. The underlying efficient DD-package may also be incorporated into automated methods for other design tasks, which rely on decision diagrams.

For the *design of Boolean components* occurring in quantum algorithms, substantial improvements in the currently established functional design flow (composed of an embedding and a synthesis step) have been achieved. However, since this flow does not consider the large degree of freedom that is actually available, the new concept of a one-pass design flow that combines these two steps has been presented—resulting in significantly more scalable synthesis that generates much cheaper circuits while being applicable to many synthesis approaches and keeping the number of required qubits at the minimum. This flow has been further improved by extending it with coding techniques. This allows realizing any non-reversible Boolean function for quantum circuits using a single ancillary qubit only (assuming that the encoded outputs can be handled by subsequent components)—significantly less than the minimum that can be achieved without proper coding.

Finally, this book has also presented significantly contributions to the *mapping of quantum circuits to NISQ devices* (utilizing IBM Q devices as representatives), a huge part of the overall compilation flow. The first exact method has been presented that generates an optimal solution for this $\mathcal{NP}$-complete problem by utilizing powerful reasoning engines. Since such exact methods do not scale well (due to the underlying complexity of the problem), a heuristic approach has also been presented (based on informed search), which significantly outperforms IBM's own solution in run-time as well as in the generated overhead. Since this approach faces its worst case when mapping a certain kind of random circuits (utilized in the verification of quantum computers) that are advocated by IBM for benchmarking quantum-circuit compilers, a dedicated approach for these so-called $SU(4)$ circuits has been described. This approach uses pre-processing, an efficient mapping strategy, as well as clever post-mapping optimizations to significantly outperform IBM's own solution by far.

The design automation methods for quantum computing presented in this book have been made publicly available as open-source implementations. Besides publications in many top-notch design automation conferences and journals, the work covered in this book also received interest from big players in the field. This is witnessed by a *Google Faculty Research Award* (for the proposed mapping simulation approach) as well as by winning the *IBM Qiskit Developer Challenge* (for mapping $SU(4)$ quantum circuits to NISQ devices).

Overall, this book shows the great potential of bringing knowledge gained from the design automation of conventional circuits and systems into the quantum realm. In the future, this development shall continue on a larger scale—eventually providing the foundation for design automation methods that accomplish for quantum computing what the design automation community realized for conventional (electronic) circuits. In this regard, a huge challenge is the interdisciplinarity of quantum computing, which requires building a bridge between the design automation community and the quantum computing community—including big players in the field—to exchange knowledge. Also here, the methods covered in this book provided first steps through an official integration into IBM's SDK Qiskit in joint work with researchers from IBM as well as an integration into Atos' Quantum Learning Machine.

References

1. S. Aaronson, L. Chen, Complexity-theoretic foundations of quantum supremacy experiments, in *Computational Complexity Conference* (2017)
2. S. Aaronson, D. Gottesman, Improved simulation of stabilizer circuits. Phys. Rev. A **70**(5), 052328 (2004)
3. A. Abdollahi, M. Pedram, Analysis and synthesis of quantum circuits by using quantum decision diagrams, in *Design, Automation and Test in Europe* (European Design and Automation Association, 2006), pp. 317–322
4. A.J. Abhari, A. Faruque, M.J. Dousti, L. Svec, O. Catu, A. Chakrabati, C.-F. Chiang, S. Vanderwilt, J. Black, F. Chong, Scaffold: quantum programming language. Technical report, Princeton Univ NJ Dept of Computer Science, 2012
5. D. Aharonov, W. Van Dam, J. Kempe, Z. Landau, S. Lloyd, O. Regev, Adiabatic quantum computation is equivalent to standard quantum computation. SIAM Rev. **50**(4), 755–787 (2008)
6. M. Amy, D. Maslov, M. Mosca, M. Roetteler, A meet-in-the-middle algorithm for fast synthesis of depth-optimal quantum circuits. IEEE Trans. Comput. Aided Des. Integr. Circuits Syst. **32**(6), 818–830 (2013)
7. S. Anders, H.J. Briegel, Fast simulation of stabilizer circuits using a graph-state representation. Phys. Rev. A **73**(2), 022334 (2006)
8. F. Arute, K. Arya, R. Babbush, D. Bacon, J.C. Bardin, R. Barends, R. Biswas, S. Boixo, F.G. Brandao, D.A. Buell, et al. Quantum supremacy using a programmable superconducting processor. Nature **574**(7779), 505–510 (2019)
9. W.C. Athas, L.J. Svensson, Reversible logic issues in adiabatic CMOS, in *Workshop on Physics and Computation* (IEEE, Piscataway, 1994), pp. 111–118
10. Atos, Atos Quantum Learning Machine, https://atos.net/en/products/quantum-learning-machine. Accessed 15 June 2019
11. A. Barenco, C.H. Bennett, R. Cleve, D.P. DiVincenzo, N. Margolus, P. Shor, T. Sleator, J.A. Smolin, H. Weinfurter, Elementary gates for quantum computation. Phys. Rev. A **52**(5), 3457 (1995)
12. R. Barends, J. Kelly, A. Megrant, A. Veitia, D. Sank, E. Jeffrey, T.C. White, J. Mutus, A.G. Fowler, B. Campbell, et al., Superconducting quantum circuits at the surface code threshold for fault tolerance. Nature **508**(7497), 500 (2014)
13. S. Beauregard, Circuit for Shor's algorithm using $2n + 3$ qubits. Quantum Inf. Comput. **3**(2), 175–185 (2003)

© Springer Nature Switzerland AG 2020
A. Zulehner, R. Wille, *Introducing Design Automation for Quantum Computing*,
https://doi.org/10.1007/978-3-030-41753-6

14. A. Biere, A. Cimatti, E. Clarke, Y. Zhu, Symbolic model checking without BDDs, in *Tools and Algorithms for the Construction and Analysis of Systems*, vol. 1579. Lecture Notes in Computer Science (Springer, Berlin, 1999), pp. 193–207

15. A. Biere, M. Heule, H. van Maaren, Handbook of Satisfiability, vol. 185 (IOS Press, Amsterdam, 2009)

16. S. Boixo, S.V. Isakov, V.N. Smelyanskiy, R. Babbush, N. Ding, Z. Jiang, M.J. Bremner, J.M. Martinis, H. Neven, Characterizing quantum supremacy in near-term devices. Nat. Phys. **14**(6), 595 (2018)

17. S. Boixo, S.V. Isakov, V.N. Smelyanskiy, H. Neven, Simulation of low-depth quantum circuits as complex undirected graphical models (2017). arXiv:1712.05384

18. A. Botea, A. Kishimoto, R. Marinescu, On the complexity of quantum circuit compilation, in *Proceedings of the Eleventh Annual Symposium on Combinatorial Search* (2018)

19. P.O. Boykin, T. Mor, M. Pulver, V. Roychowdhury, F. Vatan, A new universal and fault-tolerant quantum basis. Inf. Process. Lett. **75**(3), 101–107 (2000)

20. M. Bozzano, R. Bruttomesso, A. Cimatti, T. Junttila, P. Van Rossum, S. Schulz, R. Sebastiani, The MathSAT 3 system, in *International Conference on Automated Deduction* (2005), pp. 315–321

21. K.S. Brace, R.L. Rudell, R.E. Bryant, Efficient implementation of a BDD package, in *Design Automation Conference* (1990), pp. 40–45

22. S. Bravyi, D. Browne, P. Calpin, E. Campbell, D. Gosset, M. Howard, Simulation of quantum circuits by low-rank stabilizer decompositions (2018). arXiv:1808.00128

23. R.E. Bryant, Graph-based algorithms for Boolean function manipulation. IEEE Trans. Comput. **35**(8), 677–691 (1986)

24. R.E. Bryant, Y.-A. Chen, Verification of arithmetic circuits with binary moment diagrams, in *Design Automation Conference* (IEEE, Piscataway, 1995), pp. 535–541

25. Z. Chen, Q. Zhou, C. Xue, X. Yang, G. Guo, G. Guo, 64-qubit quantum circuit simulation. Sci. Bull. **63**(15), 964–971 (2018)

26. A.M. Childs, R. Cleve, E. Deotto, E. Farhi, S. Gutmann, D.A. Spielman, Exponential algorithmic speedup by a quantum walk, in *Symposium on Theory of Computing* (2003), pp. 59–68

27. F.T. Chong, D. Franklin, M. Martonosi, Programming languages and compiler design for realistic quantum hardware. Nature **549**(7671), 180 (2017)

28. Cirq, https://github.com/quantumlib/Cirq. Accessed 15 June 2019

29. E. Clarke, A. Biere, R. Raimi, Y. Zhu, Bounded model checking using satisfiability solving. Formal Methods Syst. Des. **19**(1), 7–34 (2001)

30. P.J. Coles, S. Eidenbenz, S. Pakin, A. Adedoyin, J. Ambrosiano, P. Anisimov, W. Casper, G. Chennupati, C. Coffrin, H. Djidjev, et al., Quantum algorithm implementations for beginners (2018). arXiv:1804.03719

31. S.A. Cook, The complexity of theorem proving procedures, in *Symposium on Theory of Computing* (1971), pp. 151–158

32. R. Courtland, Google aims for quantum computing supremacy. IEEE Spectr. **54**(6), 9–10 (2017)

33. A. Cross, The IBM Q experience and QISKit open-source quantum computing software. Bull. Am. Phys. Soc. **63** (2018)

34. A.W. Cross, L.S. Bishop, S. Sheldon, P.D. Nation, J.M. Gambetta, Validating quantum computers using randomized model circuits (2018). arXiv:1811.12926

35. A.W. Cross, L.S. Bishop, J.A. Smolin, J.M. Gambetta, Open quantum assembly language (2017). arXiv:1707.03429

36. S.A. Cuccaro, T.G. Draper, S.A. Kutin, D.P. Moulton, A new quantum ripple-carry addition circuit (2004). arXiv:quant-ph/0410184

37. M. Davis, G. Logemann, D. Loveland, A machine program for theorem proving. Comm. ACM **5**, 394–397 (1962)

38. M. Davis, H. Putnam, A computing procedure for quantification theory. J. ACM **7**, 506–521 (1960)

39. L. De Moura, N. Bjørner, Z3: an efficient SMT solver, in *International Conference on Tools and Algorithms for the Construction and Analysis of Systems* (Springer, Berlin, 2008), pp. 337–340

40. S. Debnath, N.M. Linke, C. Figgatt, K.A. Landsman, K. Wright, C. Monroe, Demonstration of a small programmable quantum computer with atomic qubits. Nature **536**(7614), 63–66 (2016)

41. D. Deutsch, Quantum theory, the church–turing principle and the universal quantum computer. Proc. R. Soc. Lond. A Math. Phys. Sci. **400**(1818), 97–117 (1985)

42. R. Drechsler, B. Becker, Ordered Kronecker functional decision diagrams-a data structure for representation and manipulation of Boolean functions. IEEE Trans. CAD Integr. Circuits Syst. **17**(10), 965–973 (1998)

43. R. Drechsler, J. Shi, G. Fey, Synthesis of fully testable circuits from BDDs. IEEE Trans. CAD Integr. Circuits Syst. **23**(3), 440–443 (2004)

44. B. Dutertre, L. De Moura, *The YICES SMT Solver* (2006). http://yices.csl.sri.com/

45. A. Ekert, R. Jozsa, Quantum computation and Shor's factoring algorithm. Rev. Mod. Phys. **68**(3), 733 (1996)

46. E. Farhi, J. Goldstone, S. Gutmann, A quantum approximate optimization algorithm (2014). arXiv:1411.4028

47. E. Farhi, J. Goldstone, S. Gutmann, M. Sipser, Quantum computation by adiabatic evolution (2000). arXiv:quant-ph/0001106

48. K. Fazel, M.A. Thornton, J.E. Rice, ESOP-based Toffoli gate cascade generation, in *IEEE Pacific Rim Conference on Communications, Computers and Signal Processing* (IEEE, Piscataway, 2007), pp. 206–209

49. R.P. Feynman, A.R. Hibbs, D.F. Styer, *Quantum Mechanics and Path Integrals* (Courier Corporation, North Chelmsford, 2010)

50. A.G. Fowler, S.J. Devitt, C. Jones, Surface code implementation of block code state distillation. Sci. Rep. **3**, 1939 (2013)

51. X. Fu, M.A. Rol, C.C. Bultink, J. van Someren, N. Khammassi, I. Ashraf, R.F.L. Vermeulen, J.C. de Sterke, W.J. Vlothuizen, R.N. Schouten, et al., A microarchitecture for a superconducting quantum processor. IEEE Micro **38**(3), 40–47 (2018)

52. J.M. Gambetta, J.M. Chow, M. Steffen, Building logical qubits in a superconducting quantum computing system. npj Quantum Inf. **3**(1), 2 (2017)

53. B. Giles, P. Selinger, Exact synthesis of multiqubit Clifford+T circuits. Phys. Rev. A **87**(3), 032332 (2013)

54. E. Goldberg, Y. Novikov, BerkMin: a fast and robust SAT-solver, in *Design, Automation and Test in Europe* (2002), pp. 142–149

55. L. Gomes, Quantum computing: both here and not here. IEEE Spectr. **55**(4), 42–47 (2018)

56. D. Gottesman, An introduction to quantum error correction and fault-tolerant quantum computation, in *Quantum Information Science and Its Contributions to Mathematics, Proceedings of Symposia in Applied Mathematics*, vol. 68 (2010), pp. 13–58

57. A.S. Green, P.L. Lumsdaine, N.J. Ross, P. Selinger, B. Valiron, Quipper: a scalable quantum programming language, in *Conference on Programming Language Design and Implementation* (2013), pp. 333–342

58. D. Große, R. Wille, G.W. Dueck, R. Drechsler, Exact multiple control Toffoli network synthesis with SAT techniques. IEEE Trans. CAD Integr. Circuits Syst. **28**(5), 703–715 (2009)

59. L.K. Grover, A fast quantum mechanical algorithm for database search, in *Symposium on the Theory of Computing* (1996), pp. 212–219

60. L.J. Guibas, R. Sedgewick, A dichromatic framework for balanced trees, in *Symposium on Foundations of Computer Science* (IEEE, Piscataway, 1978), pp. 8–21

61. P. Gupta, A. Agrawal, N.K. Jha, An algorithm for synthesis of reversible logic circuits. IEEE Trans. CAD Integr. Circuits Syst. **25**(11), 2317–2330 (2006)

62. T. Haener, M. Soeken, M. Roetteler, K.M. Svore, Quantum circuits for floating-point arithmetic, in *International Conference of Reversible Computation* (Springer, Berlin, 2018), pp. 162–174

63. T. Häner, M. Roetteler, K.M. Svore, Factoring using $2n + 2$ qubits with Toffoli based modular multiplication. Quantum Inf. Comput. **17**(7 and 8), 673–684 (2017)
64. T. Häner, D.S. Steiger, M. Smelyanskiy, M. Troyer, High performance emulation of quantum circuits, in *International Conference for High Performance Computing, Networking, Storage and Analysis* (2016), p. 74
65. P.E. Hart, N.J. Nilsson, B. Raphael, A formal basis for the heuristic determination of minimum cost paths. IEEE Trans. Syst. Sci. Cybern. **4**(2), 100–107 (1968)
66. M. Herbstritt, wld: a C++ library for decision diagrams (2004). https://ira.informatik.uni-freiburg.de/software/wld/
67. A. Hett, R. Drechsler, B. Becker, MORE: alternative implementation of BDD packages by multi-operand synthesis, in *European Design Automation Conference* (1996), pp. 164–169
68. H. Hiraishi H. Imai, BDD operations for quantum graph states, in *International Conference of Reversible Computation* (2014), pp. 216–229
69. C. Horsman, A.G. Fowler, S. Devitt, R. Van Meter, Surface code quantum computing by lattice surgery. New J. Phys. **14**(12), 123011 (2012)
70. J. Hsu, CES 2018: Intel's 49-qubit chip shoots for quantum supremacy, in *IEEE Spectrum Tech Talk* (2018)
71. D.A. Huffman, A method for the construction of minimum-redundancy codes. Proc. IRE **40**(9), 1098–1101 (1952)
72. IBM unveils world's first integrated quantum computing system for commercial use. https://newsroom.ibm.com/2019-01-08-IBM-Unveils-Worlds-First-Integrated-Quantum-Computing-System-for-Commercial-Use. Accessed 15 June 2019
73. IBM Q team, IBM Q. https://www.research.ibm.com/ibm-q/. Accessed 15 June 2019
74. IBM Q team, IBM Q 16 Melbourne backend specification v1.1.0. https://ibm.biz/qiskit-melbourne. Accessed 15 June 2019
75. IBM Q team, IBM Q 16 Rueschlikon backend specification v1.0.0. https://ibm.biz/qiskit-rueschlikon. Accessed 15 June 2019
76. IBM Q team, IBM Q 16 Rueschlikon backend specification v1.1.0. https://ibm.biz/qiskit-rueschlikon. Accessed 15 June 2019
77. IBM Q team, IBM Q 5 Tenerife backend specification v1.3.0. https://ibm.biz/qiskit-tenerife. Accessed 15 June 2019
78. IBM Q team, IBM Q 5 Yorktown backend specification v1.2.0. https://ibm.biz/qiskit-yorktown. Accessed 15 June 2019
79. IBM Q team, IBM Q Devices. https://quantumexperience.ng.bluemix.net/qx/devices. Accessed 15 June 2019
80. IBM Q team, QISKit developer challenge. https://qx-awards.mybluemix.net/#qiskitDeveloperChallengeAward. Accessed 15 June 2019
81. IonQ, IonQ: Trapped ion quantum computing. https://ionq.co. Accessed 15 June 2019
82. A. JavadiAbhari, S. Patil, D. Kudrow, J. Heckey, A. Lvov, F.T. Chong, M. Martonosi, Scaffcc: a framework for compilation and analysis of quantum computing programs, in *Computing Frontiers Conference* (2014), pp. 1:1–1:10
83. M.W. Johnson, M.H.S. Amin, S. Gildert, T. Lanting, F. Hamze, N. Dickson, R. Harris, A.J. Berkley, J. Johansson, P. Bunyk, et al., Quantum annealing with manufactured spins. Nature **473**(7346), 194 (2011)
84. T. Jones, A. Brown, I. Bush, S. Benjamin, QuEST and high performance simulation of quantum computers (2018). arXiv:1802.08032
85. J. Kelly, A preview of Bristlecone, Google's new quantum processor (2018). https://ai.googleblog.com/2018/03/a-preview-of-bristlecone-googles-new.html
86. N. Khammassi, I. Ashraf, X. Fu, C.G. Almudever, K. Bertels. QX: a high-performance quantum computer simulation platform, in *Design, Automation and Test in Europe* (IEEE, Piscataway, 2017), pp. 464–469
87. Y.-S. Kim, J.-C. Lee, O. Kwon, Y.-H. Kim, Protecting entanglement from decoherence using weak measurement and quantum measurement reversal. Nat. Phys. **8**(2), 117 (2012)

88. D.E. Knuth, The art of computer programming: binary decision diagrams (2011). https://www-cs-faculty.stanford.edu/~knuth/programs.html
89. T. Larrabee, Test pattern generation using Boolean satisfiability. IEEE Trans. CAD Integr. Circuits Syst. **11**, 4–15 (1992)
90. F. Le Gall, Powers of tensors and fast matrix multiplication, in *International Symposium on Symbolic and Algebraic Computation* (ACM, New York, 2014), pp. 296–303
91. G. Li, Y. Ding, Y. Xie, Tackling the qubit mapping problem for NISQ-era quantum devices, in *International Conference on Architectural Support for Programming Languages and Operating Systems* (2019), pp. 1001–1014
92. N.M. Linke, D. Maslov, M. Roetteler, S. Debnath, C. Figgatt, K.A. Landsman, K. Wright, C. Monroe, Experimental comparison of two quantum computing architectures. Proc. Natl. Acad. Sci. **114**(13), 3305–3310 (2017)
93. S. Malik, A.R. Wang, R.K. Brayton, A. Sangiovanni-Vincentelli, Logic verification using binary decision diagrams in a logic synthesis environment, in *International Conference on CAD* (1988), pp. 6–9
94. I.L. Markov, A. Fatima, S.V. Isakov, S. Boixo, Quantum supremacy is both closer and farther than it appears (2018). arXiv:1807.10749
95. I.L. Markov, Y. Shi, Simulating quantum computation by contracting tensor networks. SIAM J. Comput. **38**(3), 963–981 (2008)
96. J.P. Marques-Silva, K.A. Sakallah, GRASP: a search algorithm for propositional satisfiability. IEEE Trans. Comput. **48**(5), 506–521 (1999)
97. D. Maslov, G.W. Dueck, Reversible cascades with minimal garbage. IEEE Trans. CAD Integr. Circuits Syst. **23**(11), 1497–1509 (2004)
98. D. Maslov, G.W. Dueck, D. M. Miller, Techniques for the synthesis of reversible Toffoli networks. ACM Trans. Des. Autom. Electron. Syst. **12**(4), 42-es (2007)
99. K. Matsumoto, K. Amano, Representation of quantum circuits with Clifford and $\pi/8$ gates (2008). arXiv:0806.3834
100. K. McElvain, IWLS'93 benchmark set: version 4.0, in *International Workshop on Logic Synthesis* (1993)
101. Microsoft, Quantum development kit. https://www.microsoft.com/en-us/quantum/development-kit. Accessed 15 June 2019
102. D.M. Miller, R. Drechsler, Implementing a multiple-valued decision diagram package, in *International Symposium on Multi-Valued Logic* (IEEE, Piscataway, 1998), pp. 52–57
103. D.M. Miller, D. Maslov, G.W. Dueck, A transformation based algorithm for reversible logic synthesis, in *Design Automation Conference* (2003), pp. 318–323
104. D.M. Miller, M.A. Thornton, QMDD: a decision diagram structure for reversible and quantum circuits, in *International Symposium on Multi-Valued Logic* (2006), pp. 30–30
105. D.M. Miller, M.A. Thornton, D. Goodman, A decision diagram package for reversible and quantum circuit simulation, in *International Conference on Evolutionary Computation* (IEEE, Piscataway, 2006), pp. 2428–2435
106. D.M. Miller, R. Wille, Z. Sasanian, Elementary quantum gate realizations for multiple-control Toffoli gates, in *International Symposium on Multi-Valued Logic* (IEEE, Piscataway, 2011), pp. 288–293
107. S.-I. Minato, Zero-suppressed BDDs for set manipulation in combinational problems, in *Design Automation Conference* (1993), pp. 272–277
108. A. Mishchenko, M. Perkowski, Fast heuristic minimization of exclusive-sums-of-products, in *International Workshop on Applications of the Reed-Muller Expansion in Circuit Design* (2001), pp. 242–250
109. M. Mohammadi, M. Eshghi, Heuristic methods to use don't cares in automated design of reversible and quantum logic circuits. Quantum Inf. Process. **7**(4), 175–192 (2008)
110. A. Molina, J. Watrous, Revisiting the simulation of quantum Turing machines by quantum circuits. Proc. R. Soc. A **475**(2226), 20180767 (2019)
111. A. Montanaro, Quantum algorithms: an overview. npj Quantum Inf. **2**, 15023 (2016)

112. T. Monz, D. Nigg, E.A. Martinez, M.F. Brandl, P. Schindler, R. Rines, S.X. Wang, I.L. Chuang, R.Blatt, Realization of a scalable Shor algorithm. Science **351**(6277), 1068–1070 (2016)

113. M.W. Moskewicz, C.F. Madigan, Y. Zhao, L. Zhang, S. Malik, Chaff: engineering an efficient SAT solver, in *Design Automation Conference* (2001), pp. 530–535

114. M.A. Nielsen, I. Chuang, *Quantum Computation and Quantum Information* (AAPT, Maryland, 2002)

115. P. Niemann, R. Wille, D.M. Miller, M.A. Thornton, R. Drechsler. QMDDs: efficient quantum function representation and manipulation. IEEE Trans. CAD Integr. Circuits Syst. **35**(1), 86–99 (2015)

116. P. Niemann, R. Datta, R. Wille, Logic synthesis for quantum state generation, in *International Symposium on Multi-Valued Logic* (IEEE, Piscataway, 2016), pp. 247–252

117. P. Niemann, R. Wille, R. Drechsler, Efficient synthesis of quantum circuits implementing Clifford group operations, in *Asia and South Pacific Design Automation Conference* (2014), pp. 483–488

118. P. Niemann, R. Wille, R. Drechsler, Equivalence checking in multi-level quantum systems, in *International Conference of Reversible Computation* (2014), pp. 201–215

119. P. Niemann, R. Wille, R. Drechsler, Improved synthesis of Clifford+T quantum functionality, in *Design, Automation and Test in Europe* (2018), pp. 597–600

120. P. Niemann, A. Zulehner, R. Drechsler, R. Wille, Overcoming the trade-off between accuracy and compactness in decision diagrams for quantum computation. IEEE Trans. CAD Integr. Circuits Syst. (2020)

121. P. Niemann, A. Zulehner, R. Wille, R. Drechsler, Efficient construction of QMDDs for irreversible, reversible, and quantum functions, in *International Conference of Reversible Computation* (Springer, Berlin, 2017), pp. 214–231

122. E. Pednault, J.A. Gunnels, G. Nannicini, L. Horesh, T. Magerlein, E. Solomonik, R. Wisnieff, Breaking the 49-qubit barrier in the simulation of quantum circuits (2017). arXiv:1710.05867

123. E. Pednault, J.A. Gunnels, G. Nannicini, L. Horesh, R. Wisnieff, Leveraging secondary storage to simulate deep 54-qubit sycamore circuits (2019). arXiv:1910.09534

124. M.R. Prasad, A. Biere, A. Gupta, A survey of recent advances in SAT-based formal verification. Softw. Tools Technol. Transfer **7**(2), 156–173 (2005)

125. J. Preskill, Reliable quantum computers. Proc. R. Soc. Lond. A Math. Phys. Eng. Sci. **454**(1969), 385–410 (1998)

126. J. Preskill, Quantum computing in the NISQ era and beyond. Quantum **2**, 79 (2018)

127. P. Rall, D. Liang, J. Cook, W. Kretschmer, Simulation of qubit quantum circuits via Pauli propagation (2019). arXiv:1901.09070

128. A. Rauchenecker, T. Ostermann, R. Wille, Exploiting reversible logic design for implementing adiabatic circuits, in *International Conference of Integrated Circuits and Systems* (IEEE, Piscataway, 2017), pp. 264–270

129. R. Raussendorf, H.J. Briegel, A one-way quantum computer. Phys. Rev. Lett. **86**(22), 5188 (2001)

130. Rigetti, Forest SDK. https://www.rigetti.com/forest. Accessed 15 June 2019

131. M. Saeedi, R. Wille, R. Drechsler, Synthesis of quantum circuits for linear nearest neighbor architectures. Quantum Inf. Process. **10**(3), 355–377 (2011)

132. V. Samoladas, Improved BDD algorithms for the simulation of quantum circuits, in *European Symposium on Algorithms* (2008), pp. 720–731

133. M. Schwarz, M. Van den Nest, Simulating quantum circuits with sparse output distributions. Electron. Colloquium Comput. Complex. **20**, 154 (2013)

134. R. Sedgewick, in *Algorithms*, chap. 15 (Addison-Wesley, Reading, 1983), p. 199

135. E.A. Sete, W.J. Zeng, C.T. Rigetti, A functional architecture for scalable quantum computing, in *International Conference on Rebooting Computing (ICRC)* (2016), pp. 1–6

136. A. Shafaei, M. Saeedi, M. Pedram, Optimization of quantum circuits for interaction distance in linear nearest neighbor architectures, in *Design Automation Conference* (2013), pp. 41–46

137. A. Shafaei, M. Saeedi, M. Pedram, Qubit placement to minimize communication overhead in 2D quantum architectures, in *Asia and South Pacific Design Automation Conference* (2014), pp. 495–500

138. R. Shankar, *Principles of Quantum Mechanics* (Springer, Berlin, 2012)

139. V.V. Shende, I.L. Markov, On the CNOT-cost of Toffoli gates. Quantum Inf. Comput. **9**(5), 461–486 (2009)

140. V.V. Shende, A.K. Prasad, I.L. Markov, J.P. Hayes, Reversible logic circuit synthesis, in *International Conference on CAD* (2002), pp. 353–360

141. A. Shi, Recursive path-summing simulation of quantum computation (2017). arXiv:1710.09364

142. J. Shi, G. Fey, R. Drechsler, A. Glowatz, F. Hapke, J. Schloffel, PASSAT: efficient SAT-based test pattern generation for industrial circuits, in *IEEE Annual Symposium on VLSI* (2005), pp. 212–217

143. P.W. Shor, Polynomial-time algorithms for prime factorization and discrete logarithms on a quantum computer. SIAM J. Comput. **26**(5), 1484–1509 (1997)

144. M. Siraichi, V.F. Dos Santos, S. Collange, F.M.Q. Pereira, Qubit allocation, in *International Symposium on Code Generation and Optimization (CGO)* (2018), pp. 1–12

145. M. Smelyanskiy, N.P.D. Sawaya, A. Aspuru-Guzik, qHiPSTER: the quantum high performance software testing environment (2016), arXiv:1601.07195

146. A. Smith, A. Veneris, M.F. Ali, A. Viglas, Fault diagnosis and logic debugging using Boolean satisfiability. IEEE Trans. CAD Integr. Circuits Syst. **24**(10), 1606–1621 (2005)

147. M. Soeken, A. Chattopadhyay, Unlocking efficiency and scalability of reversible logic synthesis using conventional logic synthesis, in *Design Automation Conference* (2016), pp. 149:1–149:6

148. M. Soeken, G.W. Dueck, D.M. Miller, A fast symbolic transformation based algorithm for reversible logic synthesis, in *International Conference of Reversible Computation* (2016), pp. 307–321

149. M. Soeken, S. Frehse, R. Wille, R. Drechsler, RevKit: a toolkit for reversible circuit design, in *Workshop on Reversible Computation* (2010), pages 69–72. http://www.revkit.org

150. M. Soeken, T. Haener, M. Roetteler, Programming quantum computers using design automation, in *Design, Automation and Test in Europe* (IEEE, Piscataway, 2018), pp. 137–146

151. M. Soeken, M. Roetteler, N. Wiebe, G. De Micheli, Design automation and design space exploration for quantum computers, in *Design, Automation and Test in Europe* (IEEE, Piscataway, 2017), pp. 470–475

152. M. Soeken, M. Roetteler, N. Wiebe, G. De Micheli, LUT-based hierarchical reversible logic synthesis. IEEE Trans. CAD Integr. Circuits Syst. **38**(9), 1675–1688 (2018)

153. M. Soeken, L. Tague, G.W. Dueck, R. Drechsler, Ancilla-free synthesis of large reversible functions using binary decision diagrams. J. Symb. Comput. **73**, 1–26 (2016)

154. M. Soeken, R. Wille, C. Hilken, N. Przigoda, R. Drechsler, Synthesis of reversible circuits with minimal lines for large functions, in *Asia and South Pacific Design Automation Conference* (2012), pp. 85–92

155. M. Soeken, R. Wille, O. Keszocze, D.M. Miller, R. Drechsler, Embedding of large Boolean functions for reversible logic. J. Emerg. Technol. Comput. Syst. **12**(4), 41:1–41:26 (2015)

156. F. Somenzi, CUDD: CU decision diagram package release 3.0.0 (2015). http://vlsi.colorado.edu/~fabio/

157. D.S. Steiger, T. Häner, M. Troyer, ProjectQ: an open source software framework for quantum computing (2016). arXiv:1612.08091

158. T. Stornetta, F. Brewer, Implementation of an efficient parallel BDD package, in *Design Automation Conference* (1996), pp. 641–644

159. M. Van den Nest, Classical simulation of quantum computation, the Gottesman–Knill theorem, and slightly beyond. Quantum Inf. Comput. **10**(3 and 4), 258–271 (2010)

160. F. Vatan, C. Williams, Optimal quantum circuits for general two-qubit gates. Phys. Rev. A **69**(3), 032315 (2004)

161. D. Venturelli, M. Do, E. Rieffel, J. Frank, Compiling quantum circuits to realistic hardware architectures using temporal planners. Quantum Sci. Technol. **3**(2), 025004 (2018)
162. G.F. Viamontes, I.L. Markov, J.P. Hayes, Checking equivalence of quantum circuits and states, in *International Conference on CAD* (IEEE, Piscataway, 2007), pp. 69–74
163. G.F. Viamontes, I.L. Markov, J.P. Hayes, *Quantum Circuit Simulation* (Springer, Berlin, 2009)
164. G.F. Viamontes, M. Rajagopalan, I.L. Markov, J.P. Hayes, Gate-level simulation of quantum circuits, in *Asia and South Pacific Design Automation Conference* (2003), pp. 295–301
165. G. Vidal, Efficient classical simulation of slightly entangled quantum computations. Phys. Rev. Lett. **91**(14), 147902 (2003)
166. D.S. Wang, C.D. Hill, L.C.L. Hollenberg, Simulations of Shor's algorithm using matrix product states. Quantum Inf. Process. **16**(7), 176 (2017)
167. S.-A. Wang, C.-Y. Lu, I.-M. Tsai, S.-Y. Kuo, An XQDD-based verification method for quantum circuits. IEICE Trans. **91-A**(2), 584–594 (2008)
168. D. Wecker, K.M. Svore, LIQUi|>: a software design architecture and domain-specific language for quantum computing (2014). arXiv:1402.4467
169. J.D. Whitfield, J. Biamonte, A. Aspuru-Guzik, Simulation of electronic structure Hamiltonians using quantum computers. Mol. Phys. **109**(5), 735–750 (2011)
170. R. Wille, L. Burgholzer, A. Zulehner, Mapping quantum circuits to IBM QX architectures using the minimal number of SWAP and H operations, in *Design Automation Conference* (2019)
171. R. Wille, R. Drechsler, BDD-based synthesis of reversible logic for large functions, in *Design Automation Conference* (2009), pp. 270–275
172. R. Wille, G. Fey, D. Große, S. Eggersglüß, R. Drechsler. SWORD: a SAT like prover using word level information, in *VLSI of System-on-Chip* (2007), pp. 88–93
173. R. Wille, D. Große, L. Teuber, G.W. Dueck, R. Drechsler, RevLib: an online resource for reversible functions and reversible circuits, in *International Symposium on Multi-Valued Logic* (2008), pp. 220–225. http://www.revlib.org
174. R. Wille, O. Keszöcze, R. Drechsler, Determining the minimal number of lines for large reversible circuits, in *Design, Automation and Test in Europe* (2011)
175. R. Wille, O. Keszocze, M. Walter, P. Rohrs, A. Chattopadhyay, R. Drechsler, Look-ahead schemes for nearest neighbor optimization of 1D and 2D quantum circuits, in *Asia and South Pacific Design Automation Conference* (2016), pp. 292–297
176. R. Wille, A. Lye, R. Drechsler, Exact reordering of circuit lines for nearest neighbor quantum architectures. IEEE Trans. CAD Integr. Circuits Syst. **33**(12), 1818–1831 (2014)
177. R. Wille, N. Quetschlich, Y. Inoue, N. Yasuda, S.-I. Minato, Using πDDs for nearest neighbor optimization of quantum circuits, in *International Conference of Reversible Computation* (2016), pp. 181–196
178. R. Wille, M. Soeken, R. Drechsler, Reducing the number of lines in reversible circuits, in *Design Automation Conference* (2010), pp. 647–652
179. R. Wille, M. Soeken, D.M. Miller, R. Drechsler, Trading off circuit lines and gate costs in the synthesis of reversible logic. Integration **47**(2), 284–294 (2014)
180. R. Wille, M. Soeken, C. Otterstedt, R. Drechsler, Improving the mapping of reversible circuits to quantum circuits using multiple target lines, in *Asia and South Pacific Design Automation Conference* (2013), pp. 85–92
181. S. Yamashita, I.L. Markov, Fast equivalence-checking for quantum circuits, in *International Symposium on Nanoscale Architectures* (IEEE Press, New York, 2010), pp. 23–28
182. A.C.-C. Yao, Quantum circuit complexity, in *Proceedings of the 1993 IEEE 34th Annual Foundations of Computer Science* (IEEE, Piscataway, 1993), pp. 352–361
183. J.S. Zhang, S. Sinha, A. Mishchenko, R.K. Brayton, M. Chrzanowska-Jeske, Simulation and satisfiability in logic synthesis, in *International Workshop on Logic Synthesis* (2005), pp. 161–168
184. Z. Zilic, K. Radecka, A. Kazamiphur, Reversible circuit technology mapping from non-reversible specifications, in *Design, Automation and Test in Europe* (2007), pp. 558–563

185. A. Zulehner, S. Gasser, R. Wille, Exact global reordering for nearest neighbor quantum circuits using A^*, in *International Conference of Reversible Computation* (Springer, Berlin, 2017), pp. 185–201

186. A. Zulehner, S. Hillmich, R. Wille, How to efficiently handle complex values? Implementing decision diagrams for quantum computation, in *International Conference on CAD* (2019)

187. A. Zulehner, P. Niemann, R. Drechsler, R. Wille, Accuracy and compactness in decision diagrams for quantum computation, in *Design, Automation and Test in Europe* (2019)

188. A. Zulehner, P. Niemann, R. Drechsler, R. Wille, One additional qubit is enough: encoded embeddings for Boolean components in quantum circuits, in *International Symposium on Multi-Valued Logic* (2019)

189. A. Zulehner, A. Paler, R. Wille, Efficient mapping of quantum circuits to the IBM QX architectures, in *Design, Automation and Test in Europe* (IEEE, Piscataway, 2018), pp. 1135–1138

190. A. Zulehner, A. Paler, R. Wille, An efficient methodology for mapping quantum circuits to the IBM QX architectures. IEEE Trans. CAD Integr. Circuits Syst. **38**(7), 1226–1236 (2018)

191. A. Zulehner, R. Wille, Improving synthesis of reversible circuits: exploiting redundancies in paths and nodes of QMDDs, in *International Conference of Reversible Computation* (Springer, 2017), pp. 232–247

192. A. Zulehner, R. Wille, Make it reversible: efficient embedding of non-reversible functions, in *Design, Automation and Test in Europe* (European Design and Automation Association, 2017), pp. 458–463

193. A. Zulehner, R. Wille, Skipping embedding in the design of reversible circuits, in *International Symposium on Multi-Valued Logic* (IEEE, Piscataway, 2017), pp. 173–178

194. A. Zulehner, R. Wille, Taking one-to-one mappings for granted: advanced logic design of encoder circuits, in *Design, Automation and Test in Europe* (IEEE, Piscataway, 2017), pp. 818–823

195. A. Zulehner, R. Wille, Exploiting coding techniques for logic synthesis of reversible circuits, in *Asia and South Pacific Design Automation Conference* (IEEE Press, New York, 2018), pp. 670–675

196. A. Zulehner, R. Wille, One-pass design of reversible circuits: combining embedding and synthesis for reversible logic. IEEE Trans. CAD Integr. Circuits Syst. **37**(5), 996–1008 (2018)

197. A. Zulehner, R. Wille, Pushing the number of qubits below the "minimum": realizing compact Boolean components for quantum logic, in *Design, Automation and Test in Europe* (IEEE, Piscataway, 2018), pp. 1179–1182

198. A. Zulehner, R. Wille, Advanced simulation of quantum computations. IEEE Trans. CAD Integr. Circuits Syst. **38**(5), 848–859 (2018)

199. A. Zulehner, R. Wille, Compiling SU(4) quantum circuits to IBM QX architectures, in *Asia and South Pacific Design Automation Conference* (ACM, New York, 2019), pp. 185–190

200. A. Zulehner, R. Wille, Matrix-vector vs. matrix-matrix multiplication: potential in DD-based simulation of quantum computations, in *Design, Automation and Test in Europe* (European Design and Automation Association, 2019)

Index

© Springer Nature Switzerland AG 2020
A. Zulehner, R. Wille, *Introducing Design Automation for Quantum Computing*,
https://doi.org/10.1007/978-3-030-41753-6

Printed in the United States
by Baker & Taylor Publisher Services